SIX MONTHS IN MEXICO

BY
NELLIE BLY
AUTHOR OF "TEN DAYS IN A MAD HOUSE," ETC., ETC.

Updated, Revised and Edited by Nouveau Classics

©2017 Nouveau Classics

TO

GEORGE A. MADDEN,

MANAGING EDITOR

OF THE

PITTSBURG DISPATCH,

IN REMEMBRANCE OF HIS NEVER-FAILING KINDNESS.

Nellie Bly

JAN. 1st, 1888.

CONTENTS

CHAPTER I.	ADIEU TO THE UNITED STATES.	7
CHAPTER II.	EL PASO DEL NORTE.	10
CHAPTER III.	ALONG THE ROUTE.	14
CHAPTER IV.	THE CITY OF MEXICO.	19
CHAPTER V.	IN THE STREETS OF MEXICO.	21
CHAPTER VI.	HOW SUNDAY IS CELEBRATED.	29
CHAPTER VII.	A HORSEBACK RIDE OVER HISTORIC GROUNDS.	34
CHAPTER VIII.	A MEXICAN BULL-FIGHT.	40
CHAPTER IX.	THE MUSEUM AND ITS CURIOSITIES.	50
CHAPTER X.	HISTORIC TOMBS AND LONELY GRAVES.	57
CHAPTER XI.	CUPID'S WORK IN SUNNYLAND.	65
CHAPTER XII.	JOAQUIN MILLER AND COFFIN STREET.	71
CHAPTER XIII.	IN MEXICAN THEATERS.	77
CHAPTER XIV.	THE FLOATING GARDENS.	82
CHAPTER XV.	THE CASTLE OF CHAPULTEPEC.	88
CHAPTER XVI.	THE FEASTS OF THE GAMBLERS.	93
CHAPTER XVII.	FEAST OF FLOWERS AND LENTEN CELEBRATIONS.	98
CHAPTER XVIII.	GUADALUPE AND ITS ROMANTIC LEGEND.	106
CHAPTER XIX.	A DAY'S TRIP ON A STREET CAR.	113
CHAPTER XX.	WHERE MAXIMILIAN'S AMERICAN COLONY LIVED.	120
CHAPTER XXI.	A MEXICAN ARCADIA.	127
CHAPTER XXII.	THE WONDERS OF PUEBLA.	134
CHAPTER XXIII.	THE PYRAMID OF CHOLULA.	140
CHAPTER XXIV.	A FEW NOTES ABOUT MEXICAN PRESIDENTS.	146
CHAPTER XXV.	MEXICAN SOLDIERS AND THE RURALES.	151
CHAPTER XXVI.	THE PRESS OF MEXICO.	155

CHAPTER XXVII.	THE GHASTLY TALE OF DON JUAN MANUEL.	160
CHAPTER XXVIII.	A MEXICAN PARLOR.	164
CHAPTER XXIX.	LOVE AND COURTSHIP IN MEXICO.	167
CHAPTER XXX.	SCENES WITHIN MEXICAN HOMES.	171
CHAPTER XXXI.	THE ROMANCE OF THE MEXICAN PULQUE.	180
CHAPTER XXXII.	MEXCIAN MANNERS	183
CHAPTER XXXIII.	NOCHE TRISTE TREE.	186
CHAPTER XXXIV.	LITTLE NOTES OF INTEREST.	189
CHAPTER XXXV.	A FEW RECIPES FOR MEXICAN DISHES.	193
CHAPTER XXVI.	SOME MEXICAN LEGENDS.	195
CHAPTER XXXVII.	PRINCESS JOSEFA DE YTURBIDE.	198

SIX MONTHS IN MEXICO.
By NELLIE BLY.

CHAPTER I.
ADIEU TO THE UNITED STATES.

ONE wintry night I bade my few journalistic friends adieu, and, accompanied by my mother, started on my way to Mexico. Only a few months previous I had become a newspaper woman. I was too impatient to work along at the usual duties assigned women on newspapers, so I conceived the idea of going away as a correspondent.

Three days after leaving Pittsburgh we awoke one morning to find ourselves in the lap of summer. For a moment it seemed a dream. When the porter had made up our bunks the evening previous, the surrounding country had been covered with a snowy blanket. When we awoke the trees were in leaf and the balmy breeze mocked our wraps.

Three days, from dawn until dark, we sat at the end of the car inhaling the perfume of the flowers and enjoying the glorious Western sights so rich in originality. For the first time I saw women plowing while their lords and masters sat on a fence smoking. I never longed for anything so much as I did to shove those lazy fellows off.

After we got further south they had no fences. I was glad of it, because they do not look well ornamented with lazy men.

The land was so beautiful. We gazed in wonder on the cotton-fields, which looked, when moved by the breezes, like huge, foaming breakers in their mad rush for the shore. And the cowboys! I shall never forget the first real, live cowboy I saw on the plains. The train was moving at a "putting-in-time" pace, as we came up to two horsemen. They wore immense sombreros, huge spurs, and had lassos hanging to the side of their saddles. I knew they were cowboys, so, jerking off a red scarf I waved it to them.

I was not quite sure how they would respond. From the thrilling and wicked stories I had read, I fancied they might begin shooting at me as quickly as anything else. However, I was surprised and delighted to see them lift their sombreros, in a manner not excelled by a New York exquisite, and urge their horses into a mad run after us.

Such a ride! The feet of the horses never seemed to touch the ground. By this time nearly all the passengers were watching the race between horse and steam. At last we gradually left them behind. I waved my scarf sadly in farewell, and they responded with their sombreros. I never felt as much reluctance for leaving a man behind as I did to leave those cowboys.

The people at the different stopping-places looked at us with as much enjoyment as we gazed on them. They were not in the least backward about asking questions or making remarks. One woman came up to me with a smile, and said:

"Good-mornin', missis; and why are you sittin' out thar, when thar is such a nice cabin to be in?"

She could not understand how I could prefer seeing the country to sitting in a Pullman.

I had imagined that the West was a land of beef and cream; I soon learned my mistake, much to my dismay. It was almost an impossibility to get aught else than salt meat, and cream was like the stars – out of reach.

It was with regret we learned just before retiring on the evening of our third day out from St. Louis, that morning would find us in El Paso. I cannot say what hour it was when the porter called us to dress, that the train would soon reach its destination. How I did wish I had remained at home, as I rubbed my eyes and tried to dress on my knees in the berth.

"It's so dark," said my mother, as she parted the curtains. "What shall we do when we arrive?"

"Well, I'm glad it's dark, because I won't have to button my boots or comb my hair," I replied, laughing to cheer her up.

I did not feel as cheerful as I talked when we left the train. It had been our home for three days, and now we were cast forth in a strange city in the dark. The train employees were running about with their lanterns on their arms, but no one paid any attention to the drowsy passengers.

There were no cabs or cabmen, or even wheelbarrows around, and the darkness prevented us from getting a view of our surroundings.

"This has taught me a lesson. I shall fall into the arms of the first man who mentions marry to me," I said to my mother as we wended our way through freight and baggage to the waiting-room, "then I will have someone to look after me."

She looked at me with a little doubting smile, and gave my arm a reassuring pressure.

I shall never forget the sight of that waiting-room. Men, women, and children, dogs and baggage, in one promiscuous mass. The dim light of an oil-lamp fell with dreary effect on the scene. Some were sleeping, lost for awhile to all the cares of life; some were eating; some were smoking, and a group of men were passing around a bottle occasionally as they dealt out a greasy pack of cards.

It was evident that we could not wait the glimpse of dawn 'mid these surroundings. With my mother's arm still tightly clasped in mine, we again sought the outer darkness. I saw a man with a lantern on his arm, and went to him and asked directions to a hotel. He replied that they were all closed at this hour, but if I could be satisfied with a second-class house, he would conduct us to where he lived. We were only too glad for any shelter, so without one thought of where he might take us, we followed the light of his lantern as he went ahead.

It was only a short walk through the sandy streets to the place. There was one room unoccupied, and we gladly paid for it, and by the aid of a tallow candle found our way to bed.

CHAPTER II.
EL PASO DEL NORTE.

"MY dear child, do you feel rested enough?" I heard my mother ask.

"Are you up already?" I asked, turning on my side, to see her as she sat, dressed, by the open window, through which came a lazy, southern breeze.

"This hour," she replied, smiling at me; "you slept so well, I did not want to rouse you, but the morning is perfect and I want you to share its beauties with me."

The remembrance of our midnight arrival faded like a bad nightmare, and I was soon happy that I was there; only at mealtime did I long for home.

We learned that the first train we could get for Mexico would be about six o'clock in the afternoon, so we decided "to do" the town in the meanwhile.

El Paso, which is Spanish for "The Pass," is rather a lively town. It has been foretold that it will be a second Denver, so rapid is its growth. A number of different railway lines center here, and the hotels are filled the year round with health and pleasure seekers of all descriptions. While pit is always warm, yet its climate is so perfect that it benefits almost any sufferer. The hotels are quite modern, both in finish and price, and the hackdrivers on a par with those in the East.

The prices for everything are something dreadful to contemplate. The houses are mostly modern, with here and there the adobe huts which once marked this border. The courthouse and jail combined is a fine brick structure that any large city might boast of. Several very pretty little gardens brighten up the town with their green, velvety grasses and tropical plants and trees. The only objection I found to El Paso was its utter lack of grass.

The people of position are mainly those who are there for their health, or to enjoy the winter in the balmy climate, or the families of men who own ranches in Texas. The chief pleasure is driving and riding, and the display during the driving hour would put to shame many Eastern cities. The citizens are perfectly free. They speak and do and think as they please.

In our walks around, we had many proffer us information, and even ask permission to escort us to points of interest.

A woman offered to show us a place where we could get good food, and when she learned that we were leaving that evening for the City of Mexico, she urged us to get a basket of food. She said no eating-cars were run on that trip, and the eating gotten along the way would be worse than Americans could endure. We afterward felt thankful that we followed her advice.

El Paso, the American town, and El Paso del Norte (the pass to the north), the Mexican town, are separated, as New York from Brooklyn, as Pittsburgh from Allegheny. The Rio Grande, running swiftly between its low banks, its waves muddy and angry, or sometimes so low and still that one would think it had fallen asleep from too long duty, divides the two towns.

Communication is open between them by a ferryboat, which will carry you across for two and one-half cents, by hack, buggies, and saddle horses, by the Mexican Central Railway, which transports its passengers from one town to the other, and a street-car line, the only international street-car line in the world, for which it has to thank Texas capitalists.

It is not possible to find a greater contrast than these two cities form, side by side. El Paso is a progressive, lively, American town; El Paso del Norte is as far back in the Middle Ages, and as slow as it was when the first adobe hut was executed in 1680. It is rich with grass and shade trees, while El Paso is as spare of grass as a twenty-year old youth is of beard.

On that side they raise the finest grapes and sell the most exquisite wine that ever passed mortals' lips. On this side they raise vegetables and smuggle the wine over. The tobacco is pronounced unequaled, and the American pockets will carry a good deal every trip, but the Mexican is just as smart in paying visits and carrying back what can be only gotten at double the price on his side; but the Mexican custom-house officials are the least exacting in the world, and contrast as markedly with the United States' officials as the two towns do one to the other.

One of the special attractions of El Paso del Norte (barring the tobacco and wine) is a queer old stone church, which is said to be nearly 300 years old. It is low and dark and filled with peculiar paintings and funnily dressed images.

The old town seems to look with proud contempt on civilization and progress, and the little *padre* preaches against free schools and tells his poor, ignorant followers to beware of the hurry and worry of the Americans – to live as their grand- and great-grandfathers did. So, in obedience they keep on praying and attending mass, sleeping, smoking their cigarettes and eating *frijoles* (beans), lazily wondering why Americans cannot learn their wise way of enjoying life.

One can hardly believe that Americanism is separated from them only by a stream. If they were thousands of miles apart they could not be more unlike. There smallpox holds undisputed sway in the dirty streets, and, in the name of religion, vaccination is denounced; there Mexican convict-soldiers are flogged until the American's heart burns to wipe out the whole colony; there *fiestes* and Sundays are celebrated by the most inhuman cock-fights and bull-fights, and monte games of all descriptions. The bull-fights celebrated on the border are the most inhuman I have seen in all of Mexico. The horns of the *toros* (bulls) are sawed off so that they are sensitive and can make but little attempt at defense, which is attended with extreme pain. They are tortured until, sinking from pain and fatigue, they are dispatched by the butcher.

El Paso del Norte boasts of a real Mexican prison. It is a long, one-storied adobe building, situated quite handy to the main plaza, and within hearing of the merry-making of the town. There are no cells, but a few adobe rooms and a long court, where the prisoners talk together and with the guards, and count the time as it laggingly slips away. They very often play cards and smoke cigarettes. Around this prison is a line of soldiers. It is utterly impossible to cross it without detection.

Mexican keepers are not at all particular that the prisoners are fed every day. An American, at the hands of the Mexican authorities, suffers all the tortures that some preachers delight to tell us some human beings will find in the world to come.

Fire and brimstone! It is nothing to the torments of an American prisoner in a Mexican jail. Two meals, not enough to sustain life in a sick cat, must suffice him for an entire week. There are no beds, and not even water. Prisoners also have the not very comfortable knowledge that, if they get too troublesome, the keepers have a nasty habit of making them stand up and be shot in the back. The reports made out in these cases are "shot while trying to escape."

In the afternoon I exchanged my money for Mexican coin, getting a premium of twelve cents on every dollar. I had a lunch prepared, and as the shades of night began to envelop the town, we boarded the train for Mexico. After we crossed the Rio Grande our baggage was examined by the custom-house officers while we ate supper at a restaurant which, strangely enough, was run by Chinamen. This gave us a foretaste of Mexican food and price.

It was totally dark when we entered the car again, and we were quite ready to retire. There were but two other passengers in the car with us. One was a Mexican and the other a young man from Chicago.

We soon bade them good-night, and retired to our berths to sleep while the train bore us swiftly through the darkness to our destination.

CHAPTER III.
ALONG THE ROUTE.

"THIRTY minutes to dress for breakfast," was our good-morning in Mexico. We had fallen asleep the night previous as easily as a babe in its crib, with an eager anticipation of the morrow. Almost before the Pullman porter had ceased his calling, our window shades were hoisted and we were trying to see all of Mexico at one glance.

That glance brought disappointment. The land, almost as far as the eye could carry, which is a wonderful distance in the clear atmosphere of Mexico, was perfectly level. Barring the cacti, with which the country abounds, the ground was bare.

"And this is sunny Mexico, the land of the gods!" I exclaimed, in disgust.

By the time we had completed our toilet the train stopped, and we were told to get off if we wanted any breakfast. We followed our porter to a side track where, in an old freight car, was breakfast. We climbed up the high steps, paying our dollar as we entered, and found for ourselves places at the long table. It was surrounded by hungry people intent only on helping themselves. Everything was on the table, even to the coffee.

I made an effort to eat. It was impossible. My mother succeeded no better.

"Are you not glad we brought a lunch?" she asked, as her eyes met mine.

We went back to the car and managed to make a tolerable breakfast on the cold chicken and other eatables we found in our basket.

But the weather! It was simply perfect, and we soon forgot little annoyances in our enjoyment of it. We got camp chairs, and from morning until night we occupied the rear platform.

As we got further South the land grew more interesting. We gazed in wonder at the groves of cacti which raised their heads many feet in the air, and topped them off with one of the most exquisite blossoms I have ever seen.

At every station we obtained views of the Mexicans. As the train drew in, the natives, of whom the majority still retain the fashion of Adam, minus fig leaves, would rush up and gaze on the travelers in breathless wonder, and continue to look after the train as if it was the one event of their lives.

As we came to larger towns we could see armed horsemen riding at a 2:09 speed, leaving a cloud of dust in their wake, to the stations. When the train stopped they formed in a decorous line before it, and so remained until the train started again on its journey. I learned that they were a government guard. They do this so, if there is any trouble on the train or any raised at the station during their stop, they could quell it.

Hucksters and beggars constitute most of the crowd that welcomes the train. From the former we bought flowers, native fruit, eggs, goat milk, and strange Mexican food. The pear cacti, which is nursed in greenhouses in the States, grows wild on the plains to a height of twenty feet, and its great green lobes, or leaves, covered thickly with thorns, are frequently three feet in diameter. Some kinds bear a blood-red fruit, and others yellow. When gathered they are in a thorny shell. The Mexican Indians gather them and peel them and sell them to travelers for six cents a dozen. It is called "tuna," and is considered very healthy. It has a very cool and pleasing taste.

From this century-plant, or cacti, the Mexicans make their beer, which they call *pulque* (pronounced polke). It is also used by the natives to fence in their mud houses, and forms a most picturesque and impassable surrounding.

The Indians seem cleanly enough, despite all that's been said to the contrary. Along the gutters by the railroad, they could be seen washing their few bits of wearing apparel, and bathing. Many of their homes are but holes in the ground, with a straw roof. The smoke creeps out from the doorway all day, and at night the family sleep in the ashes. They seldom lie down, but sleep sitting up like a tailor, strange to say, but they never nod nor fall over.

The whirlwinds, or sand spouts, form very pretty pictures on the barren plain. They run to the height of one thousand feet, and travel along the road at a 2:04 gait, going up the mountain side as majestic as a queen. But then their race is run, for the moment they begin to descend their spell is broken, and they fall to earth again to become only common sand, and be trod by the bare, brown feet of the Indian, and the dainty hoofs of the burro.

Someone told me that when a man sees a sand spout advancing, and he does not want to be cornered by it, he shoots into it and it immediately falls. I can't say how true it is, but it seems very probable.

We had not many passengers, but what we had, excepting my mother and myself, were all men. They all carried lunch-baskets. Among them was one young Mexican gentleman who had spent several years in Europe, where he had studied the English language. He was very attentive to us, and taught me a good deal of Spanish. He had been away long enough to learn that the Mexicans had very strange ideas, and he quite enjoyed telling incidents about them.

"When the Mexican Railway was being built," he said, "wheelbarrows were imported for the native laborers. They had never seen the like before, so they filled them with earth, and, putting them on their backs, walked off to the place of deposit. It was a long time before they could be made to understand how to use them, and even then, as the Mexicans are very weak in the arms, little work could be accomplished with them.

"You would hardly believe it," he continued, "but at first the trains were regarded as the devil and the passengers as his workers. Once a settlement of natives decided to overpower the devil. They took one of their most sacred and powerful saints and placed it in the center of the track. On their knees, with great faith, they watched the advance of the train, feeling sure the saint would cause it to stop forever in its endless course. The engineer, who had not much reverence for that particular saint or saints in general, struck it with full force. That saint's reign was ended. Since then they are allowed to remain in their accustomed nooks in the churches, while the natives still have the same faith in their powers, but are not anxious to test them."

"Come, I want you to see the strangest mountain in the world," interrupted the conductor at this moment.

We followed him to the rear platform and there looked curiously at the mountain he pointed out. It rose, clear and alone, from the barren plains, like a nose on one's face. It seemed to be of brown earth, but it contained not the least sign of vegetation. It looked as high as the Brooklyn bridge from the water to top, and was about the same length, in an oblong shape. It was perfectly straight across the top.

"When this railroad was being built," he explained, "I went with a party of engineers in search of something new. Through curiosity alone, to get a good view of the land, we decided to climb that strange looking mountain. From here you cannot see the vegetation, but it is covered with a low, brown shrub. Can you imagine our surprise when we got to the top to find it was a mammoth basin? Yes, that hill holds in it the most beautiful lake I ever saw."

"That seems most wonderful!" I exclaimed, rather dubiously.

"It is not more wonderful than thousands of other places in Mexico," he replied. "In the State of Chihuahua [*] is a Laguna, in which the water is as clear as crystal. When the Americans who were superintending the work on the railway found it, they decided to have a nice bath. It had been many days since they had seen any more water than would quench their thirst – in coffee, of course. Accordingly, some dozen or more doffed their clothing and went in. Their pleasure was short-lived, for their bodies began to burn and smart, and they came out looking like scalding pigs. The water is strongly alkaline; the fish in the lake are said to be white, even to their eyes; they are unfit to eat."

I give his stories for what they are worth; I did not investigate to prove their truth.

"We do not think much of the people who come here to write us up," the conductor said one day, "for they never tell the truth. One woman who came down here to make herself famous pressed me one day for a story. I told her that out in the country the natives roasted whole hogs, heads and all, without cleaning, and so served them on the table. She jotted it down as a rare item."

"If you tell strangers untruths about your own land can you complain, then, that the same strangers misrepresent it?" asked my little mother, quietly.

The conductor flushed, and said he had not thought of it in that light before.

While yet a day's travel distant from the city of Mexico, tomatoes and strawberries were procurable. It was January. The venders were quite up to the tricks of the hucksters in the States. In a small basket they place cabbage leaves and two or three pebbles to give weight; then the top is covered with

* Pronounced Che-wa-wa.

strawberries so deftly that even the smartest purchaser thinks he is getting a bargain for twenty-five cents.

At larger towns a change for the better was noticeable in the clothing of the people. The most fashionable dress for the Mexican Indian was white muslin panteloons, twice as wide as those worn by the dudes last summer;

a *serape*, as often cotton as wool, wrapped around the shoulders; a straw sombrero, and sometimes leather sandals bound to the feet with leather cords.

The women wear loose sleeveless waists with a straight piece of cloth pinned around them for skirts, and the habitual *rebozo* wrapped about the head and holding the equally habitual baby. No difference how cold or warm the day, nor how scant the lower garments, the *serape* and *rebozo* are never laid aside, and none seem too poor to own one. Apparently, the natives do not believe much in standing, for the moment they stop walking they "hunker" down on the ground.

Never once during the three days did we think of getting tired, and it was with a little regret mingled with a desire to see more, that we knew when we awoke in the morning we would be in the City of Mexico.

CHAPTER IV.
THE CITY OF MEXICO.

"THE City of Mexico," they had called. We got off, but we saw no city. We soon learned that the train did not go further, and that we would have to take a carriage to convey us the rest of the way.

Carriages lined the entrance to the station, and the cab-men were, apparently from their actions, just like those of the States. When they procure a permit for a carriage in Mexico, it is graded and marked. A first-class carriage carries a white flag, a second-class a blue flag, and a third-class a red flag. The prices are respectively, per hour: one dollar, seventy-five cents, and fifty cents. This is meant for a protection to travelers, but the drivers are very cunning. Often at night they will remove the flag and charge double prices, but they can be punished for it.

We soon arrived at the Hotel Yturbide, and were assigned rooms by the affable clerk. The hotel was once the home of the Emperor Yturbide. It is a large building of the Mexican style. The entrance takes one into a large, open court or square. All the rooms are arranged around this court, opening out into a circle of balconies.

The lowest floor in Mexico is the cheapest. The higher up one goes the higher they find the price. The reason of this is that at the top one escapes any possible dampness, and can get the light and sun.

Our room had a red brick floor. It was large, but had no ventilation except the glass doors which opened onto the balcony. There was a little iron cot in each corner of the room, a table, washstand, and wardrobe.

It all looked so miserable – like a prisoner's cell – that I began to wish I was at home.

At dinner we had quite a time trying to understand the waiter and to make him understand us. The food we thought wretched, and, as our lunch basket was long since emptied, we felt a longing for some United States eatables.

I found we could not learn much about Mexican life by living at the hotels, so the first thing was to find someone who could speak English, and through them obtain boarding in a private family. It was rather difficult, but I succeeded, and I was glad to exchange quarters.

The City of Mexico makes many bright promises for the future. As a winter resort, as a summer resort, a city for men to accumulate fortunes; a paradise for students, for artists; a rich field for the hunter of the curious, the beautiful, and the rare. Its bright future cannot be far distant.

Already its wonders are related to the enterprising people of other climes, who are making prospective tours through the land that held cities even at the time of the discovery of America.

Mexico looks the same all over; every white street terminates at the foot of a snow-capped mountain, look which way you will. The streets are named very strangely and prove quite a torment to strangers. Every block or square is named separately.

The most prominent street is the easiest to remember, and even it is peculiar. It is called the street of San Francisco, and the first block is designated as first San Francisco, the second as second San Francisco, and so on the entire street.

One continually sees poverty and wealth side by side in Mexico, and they don't turn up their noses at each other either; the half-clad Indian has as much room on the Fifth Avenue of Mexico as the millionaire's wife – not but what that land, as this, bows to wealth.

Policemen occupy the center of the street at every termination of a block, reminding one, as they look down the streets, of so many posts. They wear white caps with numbers on, blue suits, and nickel buttons. A mace now takes the place of the sword of former days. At night they don an overcoat and hood, which makes them look just like the pictures of veiled knights. Red lanterns are left in the street where the policemen stood during the daytime, while they retire to some doorway where, it is said, they sleep as soundly as their brethren in the States.

Every hour they blow a whistle like those used by street car drivers, which is answered by those on the next posts. Thus, they know all is well. In small towns they call out the time of night, ending up with *tiempo serono* (all serene), from which the Mexican youth, with some mischievous Yankeeism, have named them *Seronos*.

CHAPTER V.
IN THE STREETS OF MEXICO.

IN Mexico, as in all other countries, the average tourist rushes to the cathedrals and places of historic note, wholly unmindful of the most intensely interesting feature the country contains – the people.

Street scenes in the City of Mexico form a brilliant and entertaining panorama, for which no charge is made. Even photographers slight this wonderful picture. If you ask for Mexican scenes they show you cathedrals, saints, cities and mountains, but never the wonderful things that are right under their eyes daily. Likewise, journalists describe this cathedral, tell you the age of that one, paint you the beauties of another, but the people, the living, moving masses that go so far toward making the population of Mexico, are passed by with scarce a mention.

It is not a clean, inviting crowd, with blue eyes and sunny hair I would take you among, but a short, heavy-set people, with almost black skins, topped off with the blackest eyes and masses of raven hair. Their lives are as dark as their skins and hair, and are invaded by no hope that through effort their lives may amount to something.

Nine women out of ten in Mexico have babies. When at a very tender age, so young as five days, the babies are completely hidden in the folds of the *rebozo* and strung to the mother's back, in close proximity to the mammoth baskets of vegetables on her head and suspended on either side of the human freight. When the babies get older their heads and feet appear, and soon they give their place to another or share their quarters, as it is no unusual sight to see a woman carry three babies at one time in her *rebozo*. They are always good. Their little coal-black eyes gaze out on what is to be their world, in solemn wonder. No baby smiles or babyish tears are ever seen on their faces. At the earliest date they are old, and appear to view life just as it is to them in all its blackness. They know no home, they have no school, and before they are able to talk they are taught to carry bundles on their heads or backs, or pack a younger member of the family while the mother carries merchandise, by which she gains a living. Their living is scarcely worth such a title. They merely exist. Thousands of them are born and raised on the streets. They have no home and were never in a bed. Going along the streets of the city late at night, you will find dark groups huddled in the shadows, which, on investigation, will turn out to be whole families gone to bed. They never lie down, but sit with their heads on their knees, and so pass the night.

When they get hungry they seek the warm side of the street and there, hunkering down, devour what they scraped up during the day, consisting of refused meats and offal boiled over a handful of charcoal. A fresh tortilla is the sweetest of sweetbreads. The men appear very kind and are frequently to be seen with the little ones tied up in their *serape*.

Groups of these at dinner would furnish rare studies for Rodgers. Several men and women will be walking along, when suddenly they will sit down in some sunny spot on the street. The women will bring fish or a lot of stuff out of a basket or poke, which is to constitute their coming meal. Meanwhile the men, who also sit flat on the street, will be looking on and accepting their portion like hungry, but well-bred, dogs.

This type of life, be it understood, is the lowest in Mexico, and connects in no way with the upper classes. The Mexicans are certainly misrepresented, most wrongfully so. They are not lazy, but just the opposite. From early dawn until late at night they can be seen filling their different occupations. The women sell papers and lottery tickets.

"See here, child," said a gray-haired lottery woman in Spanish. "Buy a ticket. A sure chance to get $10,000 for twenty-five cents." Being told that we had no faith in lotteries, she replied: "Buy one; the Blessed Virgin will bring you the money."

The laundry women, who, by the way, wash clothes whiter and iron them smoother even than the Chinese, carry the clothes home unwrapped. That is, they carry their hands high above their head, from which stream white skirts, laces, etc., furnishing a most novel and interesting sight.

"The saddest thing I ever saw," said Mr. Theo. Gestefeld, "among all the sad things in Mexico, was an incident that happened when I first arrived here.

Noticing a policeman talking to a boy around whom a crowd of dusky citizens had gathered, I, true to journalistic instinct, went up to investigate. The boy, I found, belonged to one of the many families who do odd jobs in day time for a little food, and sleep at night in some dark corner. Strung to the boy's back was a dying baby. Its little eyes were half closed in death. The crowd watched, in breathless fascination, its last slow gasps. The boy had no home to go to, he knew not where to find his parents at that hour of the day, and there he stood, while the babe died in its cradle, his *serape*. In my newspaper career I have witnessed many sad scenes, but I never saw anything so heartrending as the death of that little innocent."

Tortillas is not only one of the great Mexican dishes but one of the women's chief industries. In almost any street there can be seen women on their knees mashing corn between smooth stones, making it into a batter, and finally shaping it into round, flat cakes. They spit on their hands to keep the dough from sticking, and bake in a pan of hot grease, kept boiling by a few lumps of charcoal. Rich and poor buy and eat them, apparently unmindful of the way they are made. But it is a bread that Americans must be educated to. Many surprise the Mexicans by refusing even a taste after they see the bakers.

There are some really beautiful girls among this low class of people. Hair three quarters the length of the women, and of wonderful thickness, is common. It is often worn loose, but more frequently in two long plaits. Wigmakers find no employment here. The men wear long, heavy bangs.

There is but one thing that poor and rich indulge in with equal delight and pleasure – that is cigarette smoking. Those tottering with age down to the creeping babe are continually smoking. No spot in Mexico is sacred from them; in churches, on the railway cars, on the streets, in the theaters – everywhere are to be seen men and women – of the *elite* – smoking.

The Mexicans make unsurpassed servants. Their thievery, which is a historic complaint, must be confined to those in the suburbs, for those in houses could not be more honest. There cleanliness is something overwhelming, when one recalls the tales that have been told of the filth of the "greasers." Early in the mornings the streets, walks in the plaza, and pavements are swept as clean as anything can be, and that with brooms not as good as those children play with in the States. Put an American domestic and a Mexican servant together, even with the difference in the working implements, and the American will "get left" every time. But this cleanliness may be confined somewhat to such work as sweeping and scrubbing; it does not certainly exist in the preparation of food. Pulque, which is sucked from the mother plant into a man's mouth and thence ejected into a water-jar, is brought to town in pig-skins. The skins are filled, and then tied onto burros, or sometimes – not frequently – carried in wagons, the filled skin rolling from side to side. Never less than four filled skins are ever loaded onto a burro; oftener eight and ten. The burros are never harnessed, but go along in trains which often number fifty. Mexican politeness extends even among the lowest classes. In all their dealings they are as polite as a dancing master. The moment one is addressed off comes his poor, old, ragged hat, and bareheaded he stands until you leave him. They are not only polite to other people, but among themselves. One poor, ragged woman was trying to sell a broken knife and rusty lock at a pawnbroker's stand. "Will you buy?" she asked, plaintively. "No, senora, *gracias*" (I thank you), was the polite reply.

The police are not to be excelled. When necessary to clear a hall of an immense crowd, not a rough word is spoken. It is not: "Get out of this, now;" "Get out of here," and rough and tumble, push and rush, as it is in the States among the civilized people. With raised cap and low voice, the officer gently says in Spanish: "Gentlemen, it is not my will, but it is time to close the door. Ladies, allow me the honor to accompany you toward the door." In a very few moments the hall is empty, without noise, without trouble, just with a few polite words, among people who cannot read, who wear knives in their boots – if they have any – and carry immense revolvers strung to their belts; people who have been trained to enjoy the sight of blood, to be bloodthirsty. What a marked contrast to the educated, cultured inhabitants of the States.

Beneath all this ignorance there is a heart, as sympathetic, in its way, as that of any educated man. It is no unusual sight to see a man walk along with a coffin on his head, from which is visible the remains of some child. In an instant all the men in the gutters, on the walks, or in the doorways, have their hats off, and remain bareheaded until the sad procession is far away. The pallbearer, if such he may be called, dodges in and out among the carriages, burros and wagons, which fill the street. The drivers lift their hats, but the silent bearer – generally the father – moves along unmindful of all. Funeral cars meet with the same respect.

In passing along where a new building was being erected, attention was attracted to the body of a laborer who had fallen from the building. A white cloth covered all of the body except his sandled feet. "The Virgin rest his soul;" "Virgin Mother grant him grace," were the prayers of his kind as the policeman commanded his body to be carried away. These little scenes prove they are not brutes, that they are a little better than some intelligent people would have you believe.

The meat express does not, by any means, serve to make the meat more palatable. Generally, an old mule or horse that has reached its second childhood serves for the express. A long, iron rod, from which hooks project, is fastened on the back of the beast by means of straps. The meat is hung on these hooks, where it is exposed to the mud and dirt of the streets as well as the hair of the animal. Men with two large baskets, one in front, one behind, filled with the refuse of meat, follow nearby. If they wear trousers they have them rolled up high so the blood from the dripping meat will not soil them, but run down their bare legs and be absorbed in the sand. It is asserted that the poor do not allow this mixture in the basket to go to waste, but are as glad to get it as we are to get sirloin steak.

Men with cages of fowls, baskets of eggs and bushels of roots and charcoal, come from the mountain in droves of from twenty-five to fifty, carrying packs which average three hundred pounds.

One form of politeness here is, that when complimenting or observing anything that belongs to a native, they will reply: "It is yours." That it means nothing but politeness some are slow to learn. "My house is yours; you have but to command me," said the hotel-keeper on the day of our arrival; but he made no move to vacate. A "greeny" from the States who was working for the Mexican Central tested some beer that was on its way to the city. "That is good beer," he remarked to the express man. "*Si, senor*! It is yours," was the reply. Mr. Green was elated, and trudged off home with the keg, much to the consternation and distress of the poor express man, who was compelled to pay out of his own purse for his politeness.

"You have very handsome coffins," was remarked to a man who, probably judging from our looks since we had struck Mexican diet, thought he had found a customer, and had insisted on showing every coffin in the house, even to the handles, plates, and linings. "*Si, senorita*, they are yours." Thinking they would be an unwelcome elephant on our hands we replied with thanks, and made our exit as quickly as possible. A young Spanish gentleman who, doubtless, was employed by the express company, said, after a few moments' conversation, "The express company and myself are yours, *senorita*." We confess to the stupidity of not accepting the bonanza, with him included.

A peep into doorways shows the people at all manner of occupations. Men always use the machines. Women and men put chairs together and weave bottoms in them. They also make shoes, the finest and most artistic shoe in the world, and the cobblers can make a good shoe out of one that is so badly worn as to be useless to our grandmothers as a rod of correction. The water-carrier, *aguador*, is one of the most common objects on the street. They suspend water-jars from their heads, one in front, one back. Around their bodies are leather aprons to protect them from the water, which they get at big fountains and basins distributed throughout the city.

As a people they do not seem malicious, quarrelsome, unkind or evil-disposed. Drunkenness does not seem to be frequent, and the men, in their uncouth way, are more thoughtful of the women than many who belong to a higher class. The women, like other women, sometimes cry, doubtless for very good cause, and then the men stop to console them, patting them on the head, smoothing back their hair, gently wrapping them tighter in their *rebozo*. Late one night, when the weather was so cold, a young fellow sat on the curbstone and kept his arm around a pretty young girl. He had taken off his ragged *serape* and folded it around her shoulders, and as the tears ran down her face and she complained of the cold, he tried to comfort her, and that without a complaint of his own condition, being clad only in muslin trowsers and waist, which hung in shreds from his body.

Thus, we leave the largest part of the population of Mexico. Their condition is most touching. Homeless, poor, uncared for, untaught, they live and they die. They are worse off by thousands of times than were the slaves of the United States. Their lives are hopeless, and they know it. That they are capable of learning is proven by their work, and by their intelligence in other matters. They have a desire to gain book knowledge, or at least so says a servant who was taken from the streets, who now spends every nickel and every leisure moment in trying to learn wisdom from books.

CHAPTER VI.
HOW SUNDAY IS CELEBRATED.

"A right good land to live in
And a pleasant land to see."

EVERY day is Sunday, yet no day is Sunday, and Sunday is less Sunday than any other day in the week. Still, the Mexican way of spending Sunday is of interest to people of other climes and habits.

With the dawn of day people are to be seen wending their willing footsteps toward their church. The bells chime with their musical clang historic to Mexico, and men and women cross the threshold of churches older than the United States. Pews are unknown, and on the bare floor the millionaire is seen beside the poverty-stricken Indian; the superbly clad lady side by side with an uncombed, half naked Mexican woman. No distinction, no difference. There they kneel and offer their prayers of penitence and thanks, unmindful of rank or condition. No turning of heads to look at strange or gaze on new garments; no dividing the poor from the rich, but all with uniform thought and purpose go down on their knees to their God.

How a missionary, after one sight like this, can wish to convert them into a faith where dress and money bring attention and front pews, and where the dirty beggar is ousted by the janitor and indignantly scorned down by those in affluence, is incomprehensible.

No Mexican lady thinks it proper to wear a hat into church. She thinks it shows disgust; hence the fashion of wearing lace mantillas. In this city of rights there is nothing handsomer than a lady neatly clad in black with a mantilla gracefully wrapped around her head, under which are visible coal-black hair, sparkling eyes, and beautiful teeth.

A ragged skirt, and *rebozo* encircling a babe with its head on its mother's shoulder, fast asleep; black, silky hair which trails on the floor as she kneels, her wan, brown, pathetic face raised suppliantly in devotion, is one of the prettiest, though most common, sights in Mexico on Sunday morning.

This is the busiest day in the markets. Everything is booming, and the people, even on their way to and from church, walk in and out around the thousands of stalls, buying their marketing for dinner. Hucksters cry out their wares, and all goes as merry as a birthday party. Indians, from the mountains, are there in swarms with their marketing. The majority of stores are open; and the "second-hand" stalls on the cheap corner do the biggest business of the week.

Those who do not attend church find Mexico delightful on Sunday. In the alameda (park) three military bands, stationed in different quarters, play alternately all forenoon. The poor have a passion for music, and they crowd the park. After one band has finished, they rush to the stand of the next, where they stay until it has finished, and then move to the next. Thus, all morning they go around in a circle. The music, of which the Mexican band was a sample, is superb; even the birds are charmed. Sitting on the mammoth trees, which, grace the alameda, they add their little songs. All this, mingled with the many chimes which ring every fifteen minutes, make the scene one that is never forgotten. The rich people promenade around and enjoy themselves similar to the poor.

In the Zocalo, a plazo at the head of the main street and facing the palace and cathedral, the band plays in the evening; also on Tuesdays and Thursdays.

Maximilian planned and had made a drive which led to his castle at Chapultepec. It is 3750 feet long, wide enough to drive four, or even six teams abreast. It is planted on the east side with two rows of trees; one edging the drive, the other the walk, which is as wide as many streets. The trees are now of immense size, rendering this drive one of the handsomest, as well as most pleasant, in Mexico. Maximilian called it the Boulevarde Emperiale; but when liberty was proclaimed the name was changed to the Boulevarde of the Reform. On the same drive are handsome, nay, more, magnificent statues of Columbus, Quatemoc, and an equestrian statue of Charles IV. of wonderful size, which has also been pronounced perfect by good judges. A statue of Cortez is being erected. This paseo is the fashionable promenade and drive from five to seven P. M. every day, and specially on Sunday afternoon. The music stands are occupied, and no vacant benches are to be found.

Those who call the Mexicans "greasers," and think them a dumb, ignorant class, should see the paseo on Sunday: tally-ho coaches, elegant dog-carts, English gigs, handsome coupes and carriages, drawn by the finest studs, are a common sight. Pittsburg, on this line, is nowhere in comparison. Cream horses, with silver manes and tails, like those so valued in other cities, are a common kind here. The most fashionable horse has mane and tail "bobbed." It might be added this style prevails to a great, very great extent among all animals. Cats and dogs appear minus ears and tails. Pets of every kind are much in demand. Ladies carry lap dogs, and gentlemen have chained to them blooded dogs of mammoth size. The poor Mexican will have his tame birds; even roosters are stylish pets. "Mary had a little lamb" is respected too much here to be called "chestnut." The favorite pets of children are fleecy lambs, which, with bells and ribbons about their necks, accompany the children on their daily airing.

Mexico, while in the land of churches, would be rightly called the city of high heels, hats, powder and canes. Every gentleman wears a silk hat and swings a "nobby" cane. There are but two styles of hats – the tile hat and the sombrero. Every woman powders – lays it on in chunks – and wears the high heels known as the French opera heel. The style extends even to the men. One of the easiest ways to distinguish foreigners from natives is to look at their feet. The native has a neat shoe, with heels from two inches up, while the foreigner has a broad shoe and low heel. These people certainly possess the smallest hands and feet of any nation in the world. Ladies wear fancy shoes entirely – beaded, bronzed, colored leather, etc. A common, black leather shoe, such as worn by women in the Slates, is an unsalable article. Yet it is nothing strange to see a lady, clad in silk or velvet, lift her dress to cross a street or enter a carriage, and display a satin shoe of exquisite make and above it the hosiery of Eve. In fact, very few women ever wear stockings at all.

This city is a second Paris in the matter of dress among the *elite*. The styles and materials are badly Parisian, and Americans who come here expecting to see poorly-dressed people are disappointed. Like people in the sister Republic, the Mexicans judge persons by their dress. It is the dress first and the man after.

On Sundays the streets and parks are thronged with men and women selling ice cream, pulque, candies, cakes, and other dainties. They carry their stock on their heads while moving, and when they stop they set it on a tripod, which they carry in their arms.

The flower sellers are always women, some of whom look quite picturesque in their gay-colored costumes. All the flowers are elegant, and are arranged in bouquets to suit either ladies or gentlemen.

Bull fights take no little part in the Sunday list of amusements, where the poor and rich mingle freely. Theaters have matinees and evening performances, and everything takes on a holiday look, and everybody appears happy and good-humored. This is nothing new in Mexico, however, for the most unusual sight is a fight or quarrel. These are left to the numerous dogs which belong to the city, and even they do little of it.

Riding horseback is a favorite pastime. Ladies only ride in the forenoon, as custom prevents them from indulging in the saddle after one o'clock. Gentlemen, however, ride mornings and evenings. Among them are to be found the most graceful and daring riders in the world. Their outfits are gorgeous; true Mexican saddle trimmed with gold and silver, graceful flaps of the finest fur on bridles finished with numberless silver chains. The riders are superb in yellow goatskin suits, ornamented with silver horse shoes, whips, spurs, etc., with silver braid on the short coat. A handsome sombrero, finished in silver, with silver monogram of the owner, revolvers, and proud, fiery, high-stepping horse completed the picture. The ladies' habits are similar to those now in the States, except the fine sombrero which replaces the ugly, ungraceful high silk hats.

All day Sunday is like a pleasant Fourth of July, but after eight o'clock the carriages become scarcer and scarcer, the people go to the theaters and to their homes, the poor seek a soft flagstone, where they repose for the night, and by nine o'clock the streets make one think of a deserted city.

Mexicans do not go half way in the matter of style. At one o'clock Sunday afternoons policemen in fancy uniforms, mounted on handsome horses, equipped with guns and lassoes, ride down the Boulevard. They are stationed in the center of the drive one hundred yards apart, every alternate horse's head in the same direction. There they remain, like statues, the entire afternoon. Sunday is a favorite day for funerals and change of residence. Men with wardrobes, pianos, etc., on their backs are seen trotting up and down the streets like our moving wagons on the first day of April. They mean well by work on Sunday, but it would appear awful to some of our good people at home. There is this advantage, at least: they have something better to do than to congregate in back-door saloons or loaf on the streets.

CHAPTER VII.
A HORSEBACK RIDE OVER HISTORIC GROUNDS.

A SUNDAY in Mexico is one long feast of champagne, without a headache the next day. When the first streaks of dawn appear in the east people bob out from this street and that, hostlers hurry horses off to private residences, gay riders whirl by as if eager to catch the shades of night as they are sinking in the west, and by 6:30 it looks as if all Mexico was on horseback. Ladies wear beautiful costumes, dark habits, short skirts, silver and gold buttons, and broad sombreros. Men display greater variety of costumes; some wear yellow buckskin suits trimmed with gold or silver, others have a drab skin suit artistically trimmed, still others wear light cloth suits and high boots, buttoned at the side, and reaching the knee. A belt holding a revolver, and a Mexican saddle to which is fastened a sword complete this beautiful riding suit. And then what riders! It is the poetry of motion; they are as but part of the perfect horse they ride. Take the beautiful horses, artistic outfit, grand eyes glancing at you from beneath a pretty sombrero, and you have a Mexican scene which is irresistible. Even Americans are a thousand times handsomer when they don this outfit, and it is safe to wager that if the men in the States would adopt the Mexican riding-suit, there would not be a single man left after a two months' trial.

After searching the whole city over we at last found a woman we knew, who owned a habit. "Certainly, you may have it, with great pleasure," and we thought what an angel she was until the time we needed it, when she sent a reply: "My riding-dress is, as I told you, at your service any day in the week but Sunday. I am surprised that you find need of it on that blessed day." That evening on going to a house for dinner we found her there, dressed to the height of fashion, discussing the people who had attended church in the morning and telling what a lovely drive she had on the paseo in the afternoon. She is a missionary.

However, as the sun was creeping up trying to catch night unawares, I mounted a horse, clad in a unique and original costume, to say the very least, which the gallant young men, however, pronounced odd and pretty, and wanted to know if it was the style of the States. The boulevard of the Reform looked as cool and sweet as a May morning in the country, and finer than a circus parade with the hundreds of horsemen going either way. "*Vamos?*" (Let us go). "*Con mucho gusto*" (with much pleasure), was our reply, and away flew our willing steeds, bearing us soon to the paradise of Mexico – Chapultepec.

Greeting the guards at the gate, we entered, riding under trees which sheltered Montezuma and his people, Cortes and his soldiers, poor Maximilian and Charlotta, where Mexican cadets laid down their lives in defense of their country, where the last battle was fought with the Americans, and where now is being prepared the future home of President Diaz. Around the castle and through the grounds we at last emerged at the opposite side. Here a scene worthy of an artist's brush was found. In a small adobe house, faced in front by a porch, were half-clad Mexicans dealing out coffee and pulque to the horsemen who surrounded the place. One had even ridden into the house. Awaiting our turn, we viewed the scene. On our left were mounted and unmounted uniformed soldiers guarding one of the gates to Chapultepec. At our back were trains of loaded burros, about 200, on their way to market in the city. They stood around and about the old aqueduct, the picture of patience. Some few had lain down with their burdens and had to be assisted to their feet by their masters. Numerous little charcoal fires, above which were suspended pans and kettles, were being fanned by enterprising peons, who had started this restaurant to make a few pennies from their fellowmen. One fellow cut all kinds of meat, on a flat stone, into little pieces, which he deposited together in a kettle of boiling water, and picking them out again with a long stick sold them, half-cooked, to the waiting people. Some women were busily knitting, weaving baskets, etc., as they waited for this dainty repast. At last our turn came, and we turned our back on the outdoor restaurants while we endeavored to swallow a little bit of the miserable stuff they called coffee. As we started we saw the people adjust the burdens to their backs, take up their long walking-poles, and start their burros toward the city. They had feasted and were now ready to continue their journey.

Leaping a ditch, we left the highway and traveled through the fields, stopping to gather a few pepper berries with which to decorate ourselves, admiring the many-colored birds flitting from tree to tree. Another ditch, which the horses cleared beautifully, was left behind and we were once again on a highway, with dust about a foot deep, which made horses cough as well as their riders. "This is bad," one of the gentlemen managed to say at last. We were only able to give a sympathetic grunt and then had to gasp fifteen minutes before we could regain our breath. "There is a hacienda near where we will get a drink and change roads. Vamos." Off we went, leaving the dust behind, and were soon in the shaded drive leading to the hacienda.

Here, at Huischal, we soon forgot the scorching sun and blinding dust and gave ourselves up to the pleasure of the moment, watching the ever-picturesque people gathered in groups beneath the shade. Under the trees were droves of horses, which were taken two by two, and led into a large walled pond. A peon walked on the wall, holding the bridle of the tethered horses, who swam from one end to the other, covered all but the head. After the bath the horses were rubbed well until they glistened like satin.

Climbing the hill, we passed all kinds of Indians and huts. There were homes built entirely of the maguey plant, where straw mats served for beds. The people were all awake and engaged in various occupations; some women were washing, some were making their toilet – combing their hair with the same kind of brush they scrub with, and washing their bodies with a porous soapstone common to the country. Very few of the children had any clothing at all, but happiness reigned supreme. We passed several plain wooden crosses with inscriptions on them, asking travelers to pray for the deceased's soul. It brought forcibly to mind Byron's "Childe Harold."

Quite on the top of the hill, and facing Chapultepec, gleams a marble monument erected in honor of the Mexicans killed while defending Casa de Mata (the house of the dead) and El Molino del Rey (the mill of the king). The Americans discovered, while encamped near here, that cannon, etc., were being manufactured at El Molino, so they decided to storm the place; they found the work more difficult than they expected. The Mexicans were fighting for a country they loved, and for which they had been compelled to fight for generations. Their walls were strong, but at last they gave way before the heavy artillery of the Americans, and their dead covered the battlefield. Casa de Mata is now a garrison, and the soldiers march back and forth with sad faces. El Molino del Rey now furnishes flour for the city. It shows no trace of the assault. Nearby is a foundry for the manufacture of guns and munitions.

The city of the dead, Dolores, lies to the back of the mill. Funeral cars and draped street cars were just returning from the cemetery, and as the people are not allowed to ride or drive along this carway, we crossed into a plantation of pulque plant. It is a resentful thing, and a whole army in itself. It ran its sharp prongs into the legs of the men, endeavored to pull the skirts off the women, and played spurs on the horses; but we finally emerged at the entrance of the cemetery, alive, but wiser from our experience.

Mexican cemeteries have a certain peculiar beauty, and yet they are ugly. No one is allowed to ride or drive through; coffins are carried in and everybody is compelled to walk. Beautiful trees are cultivated, even the apple and the peach being reared for ornament. The walks are laid out nicely. Spruce trees are trained to form an arbor for long distances. Where they are divided or meet another walk, flowing fountains with large basins and statues grace the spot. One statue, which looked rather singular, was apparently carved out of wood. It represented a man with flowing locks and beard, clad in a long gown and holding in one hand a round ball. Time had its hand on heavily, and the wood was seamed and browned. Altogether it was a disreputable-looking thing. The keeper said it represented Christ with the world in his hand. Not a sprig of grass is permitted to grow in any of the graveyards, and they are swept as clean as our grandmother's backyard used to be.

Men were busy digging graves, and new ones were completely hidden by fresh flowers, and the flowers on others were withered and dead, as if the one so lately buried was already forgotten. The monuments are quite fine. Some have little altars on which candles are lighted on certain days. The prevailing style of marble shaft is coffin shaped. Some graves have miniature summerhouses built over them, the framework covered with Spanish moss. The effect is beautiful. The poor have only black and white wooden crosses to mark their ashes. One family had built a cave, formed of volcanic stone, over the grave, the effect being quite pretty and unique.

After partaking of refreshments at a long, low building, just outside the cemetery gate, we rode across the country and into Tacubaya, an ancient city once the home of Montezuma's favorite chief, where the American soldiers were encamped, now the home of Mexican millionaires, the site of the feast of the gamblers, and the prettiest village in Mexico. The gambling feast has ended and the town has been restored to its usual quietness. In the center plaza a band was holding forth, as is the custom in every Mexican village on Sunday mornings. People had gathered in sun and shade listening. The markets were in full blast; the thousands of luscious fruits looking fresh and inviting as they were spread on the ground awaiting buyers. The native ware was so peculiar and the "merchant" – half-dressed, brown and pleasant – was more than we could resist, so buying two small cream jugs, made after the style in vogue fifty years ago, we paid him two reales (fifty cents) and departed, leaving him happy.

Once again, the willing horses climbed the hill, and reaching the summit we inspected the waterworks which have so faithfully supplied the city for years. A weather-beaten frame house hid the well or spring that has given such a generous supply. A wooden wheel as large as the house itself, moved slowly, as if age and rheumatism had stiffened its joints. The water flowed gently through an open trench into another building, whence it rushed, white, foaming and sparkling, into the ground, leaving only high brick air-pipes to mark its course to the aqueduct.

By the side of the trench a woman was doing her washing, and two little lads, with poles across their shoulders and buckets suspended from either end, were carrying water to the houses down in the valley. An old cow with curly horns gazed at us in astonishment as we invaded her private meadow to get a view of a paper mill, which is built in the shape of an old English castle, down in a deep ravine in a nest of lovely green trees. The old cow had evidently come to the conclusion, after deliberate reasoning, that we were intruding, and she charged our horses in a first-class "toro" style. There were no *capeadores* to attract her attention, no *bourladeras* for us to hide behind, so we thought it best to fly, which we did with a Maud S speed. I did not mention I had lost my hat in the retreat until we were over the trench, and one of the young men gallantly started to recover it, against the protestations of the entire crowd. We expected to see him killed, but the cow stood watching him as he dismounted for the feminine headgear, gesticulating with head and tail and beating the earth with her fore legs. Remounting, he saluted her, then putting spurs to his horse he cleared the ditch, leaving the baffled and angry cow on the other side.

La Castaneda, the great pleasure-garden of the Mexicans, was next visited. Beautiful flowers, shrubbery and marble statues grace the well-kept resort. Neat little benches, cunning little vine-draped nooks, sprinkling-fountains, secluded dancing-stands, deep bathing-basins, are a few of the many attractions. Shaded walks and twisting stairways would always bring us to some new beauty. Music and dancing are always held here every afternoon, and although it was nearly noon they had not even so much as a cracker in the house. In Mexico nothing in the line of edibles is kept in the house overnight.

At Mixcoac we visited the famous flower gardens, and viewed the site where the American soldiers were garrisoned during the war. The Mexicans have found a new thing – a pun, and they are enjoying it heartily. It is not very brilliant or very funny, but it is traveling over the city, and every person has to repeat it to you. An American wanted to see Mixcoac – pronounced "quack." The conductor failed to let him out at the place, and turning to the Mexicans he said: "We have mis-t-quack." But it was funnier still to an American who was being showed around by a Mexican who spoke very little English. "I will take you to see Mis-quack," said the Mexican. The American expressed his pleasure and willingness. "This is all Mis-quack," said the Mexican, pointing around the entire town. "Indeed," ejaculated the astonished tourist; "Miss Quack must be very wealthy."

Down the dusty road we came, passing natives shooting the pretty birds just for the fun of the thing. All other riders had disappeared, and people looked at us from beneath the shade in amazement, and even we felt a little tired and heated after a thirty-mile ride. We reached home at one o'clock. Since then I have been wearing blisters on my cheeks and nose, and making frequent applications with the powder rag of the literary widow and old-maid artist who room across the way.

CHAPTER VIII.
A MEXICAN BULL-FIGHT.

Mexicans are always manana until it comes to bull-fights and love affairs. To know a Mexican in daily life is to witness his courtesy, his politeness, gentleness; and then see him at a bull-fight, and he is hardly recognizable. He is literally transformed. His gentleness and "manana" have disappeared; his eyes flash, his cheeks flush – in fact, he is the picture of "diabolic animation." It is all "hoy" to-day with him. Even the Spanish lady of ease and high heels forgets her mannerisms and appears like some painted heathen jubilant over the roasting of a zealous missionary.

There have been some very good bull-fights lately in the suburbs, for fighting is prohibited within a certain distance of the city. When they say a good bull-fight, it means that the bulls have been ferocious and many horses and men have been killed.

It is safe to say that the majority of Americans who visit Mexico do like the natives, even on the first Sunday; attend divine service in the morning, a bull-fight in the afternoon and theater in the evening. But it is with regret that I say that many Americans who are residents of the city now are as passionately fond of the national inhuman sport as a native who has been reared up to it. Some never miss a fight, and their American voice outstrips the Mexican in the shouts of "bravo" at the bloody thrusts. Yet there are tourists who cannot outsit one performance, and have no desire to attend a second. While we Americans cry "brutal" against the national amusement, they in return cry "brutal" to our prize-fights, in which they see nothing to admire, and a dog-fight is beneath their contempt.

"Your humane societies would prevent bull-fights in the States," said a Spanish gentleman; "your people would cry out against them. Yet they have strong men trying to pound one another to death, and the people clamor for admission to see the law kill men and women, while in health and youth, because of some deed done in the flesh. Yes, they witness and allow such inhuman treatment to a fellow mortal and turn around and affect holy horror at us for taking out of the world a few old horses and furnishing beef for the poor."

Read of glorious bull-fights and then witness one, and the scene is entirely changed. The day of their glory has departed. When Maximilian graced the country with his presence the fights were indeed fitted for royal sight. The costumes were of the costliest material; the horses were of the best blood and breed, and the bulls regular roaring Texans, which needed no second sight of a red capa to raise their feverish ire. No fight cost less than $5,000.

Now all is different. Maximilian lies in a grave to which a treacherous bullet consigned him; Carlotta, still what that bullet made her, a raving lunatic and a widow. Men of low degree are permitted to grace the fights, which are but miserable shadows, a farce of the former royal days.

The National – a narrow gauge – and the Mexican Central, run special trains consisting of twenty and twenty-five cars, first, second, and third-class, to the fights every half hour. Tickets are sold during the week, which include railroad fare, admission to grounds and seat. Long before the time for leaving, carriages pull up to the stations and blooming senoras, fair senoritas, handsome senors and delicate, lovely children, dressed in the height of wealth, and fashion, enter the railway coach and proceed to make themselves comfortable for the half-hour or hour's ride which is to bring them to their destination. Bands march up and are disposed of in the coaches, and last comes a troop of soldiers, clad in buckskin suits, elaborately trimmed with silver ornaments, yard wide sombreros, and armed with gun, revolver, sword, dagger, mace, and lasso, which they have no hesitation in using in quite a characteristic manner, asking no questions, expecting no information, performing their duties fatally.

They are the "daisies" of Mexico, and in appreciation of which they are sent to grace every bull fight! They are the best paid soldiers in the republic, receiving $1 a day, while the highest salary paid to any of the others is twenty-five cents daily, out of which they provide their own wearing apparel and food. The same "daisies" were all outlaws, bandits, fierce and uncontrollable. Their many deeds, always done in the name of the law, are fearful to relate, so the present president thought it policy to engage their services. They ride handsome horses, furnished by the government, and are said to be the most faithful, reliable men in the employ of the republic. Their only fault is killing without asking questions, for which they go scot-free without even so much as a rebuke. The "daisies" have some of the finest specimens of manhood in Mexico, and number in their list some handsome, open-faced, youthful boys. They can maintain order among 6,000 people filled with pulque without uttering one word. Their presence is sufficient.

On speeds the train. Above the din arises the musical sound of a strange language. A view from the window exhibits some of Mexico's most beautiful scenery. Now we pass beautiful farms, magnificent artificial lakes covered with wild duck, which would delight the heart of our American hunters, as they arise in dark clouds on the approach of the train, and move off to a more secluded spot; beautiful fields of grain, and acres and acres of pulque plant, quaint huts, picturesque, historic churches, ancient monasteries and convents, now used for other purposes, all surrounded by snow-capped mountains. For miles we keep our eyes on the strangest and grandest mountain in Mexico, the White Lady, or the Sleeping Virgin. It deserves chapters of description and praise, but feeling our inability to do it justice we shall confine ourselves to a brief remark.

Outlined against a blue sky, only such skies as are habitual to Italy and Mexico, is a snow-topped mountain in form of a woman lying on a straight cot; on the head is a snow band, such as worn by Sisters of Mercy. The arms are folded peacefully on the breast, and the snow garments fall in graceful folds over the feet. There she lies and has lain for centuries in perfect outline and peaceful repose. Even as we look the clouds play fantastically about the beauteous form. Now they cover her body like a dark shroud. Again, they drape her cot like a pall, then rise in a threatening attitude above her fair head, but undisturbed she lies there with hands ever folded above the quiet heart, proudly indifferent to storm or shine, clad in her pure snowy garments, truly the most beauteous sight in Mexico. With a sigh we at last leave her behind and are rudely brought to earth by the announcement that we have reached our destination.

The bull ring resembles somewhat a race course; the highest row is covered and called boxes. They are divided into small squares, which are meant to hold six but are crowded with four. Miserable chairs without backs are the comfortable seats. Below is the amphitheater, arranged exactly like circus seats. Different prices are charged and the cheapest is the sunny side, where all the poor sit. A fence painted in the national colors – red, green and white – of some six feet in height, encloses the ring. Three band-stands, equal distances apart, are filled with brilliantly uniformed musicians.

The judge is appointed by the municipality, but the fighters have a right to refuse to fight under one judge whom they think will compel them to take unnecessary risks with a treacherous bull, for a judge once chosen his commands are law, and no excuse will be accepted for not obeying, but a fine deducted from the fighter's salary, and he loses cast with the audience. The judge is in a box in the center of the shady side; with him is some prominent man, for every fight must be honored with the presence of some "high-toned" individual, while behind stands the bugler, a small boy in gay uniform, with a bugle slung to his side, by which he conveys the judge's whispered commands to the fighters in the ring.

Below the judge hangs a row of banderillas. They are wooden sticks about two feet long with a barbed spear of steel in the end, which are stuck in the bull to gore him to madness. They are always gayly decorated with tinsel and gaudy streamers of the national colors. Sometimes firecrackers are ingeniously inserted, which go off when the banderilla is deftly fastened in the beast's quivering flesh.

The bands play alternately lively airs, the audience for once find no charms in the music and forget to murmur manana, but soon begin to cry "El toro! El toro!" (The bull! the bull!)

The judge nods to the bugler, and as he trumpets forth the gate is swung open and the grand entry is made. First comes "El Capitan" or matador, chief of the ring, and the men who kill the bull with a sword. Next eight capeadores, whose duty consists in maddening the bull and urging it to fight by flinging gay-colored *capas* or capes in its face. Two picadores, who are armed with long poles, called picas, in the end of which are sharp steel spears which they fight the bull with. After, come the lazadores, dressed in buckskin suits, elaborately trimmed with silver ornaments and broad, expensive sombreros. They ride fine horses, and do some very pretty work at lassoing. Three mules abreast, with gay plumes in their heads, and a man with a monstrous wheelbarrow of ancient make, close up the rear. All range before the judge and make a profound bow, after which the mules and wheelbarrow disappear.

The dresses of the fighters are very gorgeous: satin knee-breeches and sack coat of beautiful colors, and highly ornamented, beaded, etc. On the arm is carried the *capa*, a satin cape, the color of the suits, and little rough caps, tied under the chin, grace the head. At the back of the head is fastened false hair, like a Chinaman's, familiarly known as "pig tail." Two gayly painted clowns, who, unlike those in the States, never have anything to say, are always necessary to complete the company in the ring.

Again, the bugle sounds, the band strikes out in all its might, the people rise to their feet and cry "*El toro*," the fighters form a semicircle around a door, el capitan draws a bolt, flings it open, and as the bull springs forth from his dark and narrow cell a man perched above sticks two *banderillas* into his neck to madden him. With a snort of rage, he rushes for the *capas*. As they are flirted before his eyes, he tramples them under his hoofs, and the *capeadors* escape behind the *bourladera*, a partition, six feet wide, placed in the arena at four places equally distant.

At the trumpet sound a banderilla runs out waving the banderillas above his head. He faces the maddened bull with a calm smile. The bull paws the ground, lowers his head, and with a bellow of rage makes for his victim. Your eyes are glued to the spot. It is so silent you can hear your heart throb.

There can be no possible escape for the man. But just as you think the bull will lift him on his horns you see the two banderillas stuck one in either side of the neck, and the man springs safely over the lowered head and murderous horns of the infuriated animal, as it rushes forward to find the victim has escaped. The audience shout "bravo," and wave their serapes, sombreros and clap their hands. The bull roars with pain, and the banderillas toss about in the lacerated flesh, from which the blood pours in crimson streams. "Poor beast! what a shame," we think, and even then, the order is given for the picador to attack the bull.

The horse on which the picador is mounted is bought only to be killed. It is an old beast whose days of beauty and usefulness are over; $2 or $4 buys him for the purpose. Sometimes he is hardly able to walk into the ring. First, the brute is blindfolded with a leather band, and a leather apron is fastened around his neck in pretense of saving him from being gored.

The picador guides the blinded horse to face the bull. Capas are flung before the bull tauntingly. The picador drives the pica into the beast and it vents its pain on the horse. Blood pours from the wound; trembling the horse stands, unable to see what has wounded it. Again, they coax the bull to charge, and place the horse so that the murderous horns will disembowel it. Down goes the blinded beast, and the capeadores flaunt their capas at the bull while the picadore gets off the dying animal, which is lassoed and dragged from the ring. Another horse is brought in, and the same work is gone over until the horse is killed.

Every bull is allowed to kill two horses, and then the people shout "*Muerie! muerie!*" (Kill the bull.) The judge gives the command and the matador bows to the judge, and then teases the bull with his red *capa*. The laws prohibit a fighter to strike a bull until it first charges, and the bull has the chance of three charges at the matador before he dares to strike. The bull never appears to see the man by his side, but furiously fights the red *capa* held before him. *El capitan* then plunges the sword into the neck between the shoulders and through to the heart, if deftly done, after which the bull staggers, protrudes its tongue, tries to find a door for escape, stumbles and dies. Again, the people shout, and the matador, as he makes his bow to the judge, is thrown money, cigars, fruit, flowers and other favors. Men fling in their $50 and $100 *sombreros*, and consider it a great honor when he picks them up and tosses them back. During all this the three mules are brought in. At the sight of the dead bull they plunge and tear, but are finally hitched to it. The clowns jump on the dead beast, and it is hauled from the ring.

When the bull is tame and, though tortured on all sides, still refuses to gore the horse, the people hiss and shout "*lazadore,*" until the judge gives the command for the brute, that is more humane than its tormentors, to be removed and replaced by one that will sate their feverish desire for blood. Now is the time for the *lazadores* to get in some pretty work. The space is small and cramped, but with a deftness that is bewildering they throw the loop over the horns. The knowing horse dodges, the bull loses his balance and the horse gives a sudden jerk, throwing the bull on the ground. He is then allowed to arise and is started around the ring at a merry gallop, while the second *lazadore* exhibits great skill in lassoing the feet, front and back, of the running beast.

The bull, after being thrown, realizes he is at their mercy, and lies passive; or trembling with fear and pain, while the brutal clowns spring astride the prostrated beast, and with no gentle hand tear the banderillas from the quivering flesh, which, still warm and dripping with blood, are sold as trophies at one and two dollars each. Then the butcher steps forth and with a sharp knife cuts the spinal cord, and the beast is done for. When a bull refuses to fight before he is cut, except for wounds from the pica and banderillas, the people cry in Spanish, "He is a weak woman," until the judge orders his removal. It is difficult work, and affords much fun for the Mexicans, for the bull must be forced back into the dark cell whence he came.

One fight consists of four bulls and as many old horses as they can be compelled to kill. A bull is not considered much unless he can kill, at the very least, two horses. The poor horses are very seldom killed instantly. When wounded so that it is impossible for them to walk, they are dragged from the ring and left in a vacant field, where they die that night or the following day, as the Mexicans do not consider them worth a bullet. The bull finds more mercy. If not killed outright by the matador, a butcher finishes the work, and ends the misery. When stabbed fatally he often staggers along the fence, as though in hopes of finding an exit. The cruel spectators are not satisfied that he is dying, and allow him some little mercy, but stab his wounded flesh, tear open his death wound, twist his tail, do all in their power to enhance his sufferings until he falls dead. One would suppose the heated, tortured, wounded beef would be of no account, but such is not the case. Before many hours, after taken from the scene of its death, the beef is being sold to the people, who buy it without the least hesitancy or disgust, even boasting that they eat of the bull that killed so many horses, and if it happened to kill a man it is considered an honor to eat of it. This makes an American want little beef, and that little covered with red pepper to kill the taste. When seated opposite the entrance gate one has full view of the butcher at work. The hide is taken off the toro immediately, and it is dissected. Then they commence on the horses, but they claim the horses' flesh is not sold for beef.

At some fights the spectators are favored with a performer, who allows the maddened toro to attack him, when, by the aid of a long pole, he jumps clear over it. This is a dangerous and, many times, a fatal leap, but is a favorite sight of the people.

After the fight comes the *toro embolado*. A bull with balls on its horns is led in. All the paid fighters leave the ring and any one among the spectators who has a desire to try the sport can do so. The number is not few, and the sight is really funny. They wave their *serapes* at the bull, who, in return, often tosses them on his horns. The *lazadores* prevent him from trampling them, and it is very seldom any one is killed, though broken arms and ribs are no unusual thing. This is the proudest day of the Mexican's life when he gains access to the bull ring and can exhibit to people his activity and daring. The most risky amateur is then given a position as fighter, a position he considers greater than the presidency of the United States, and for which he would not exchange.

The government charges a license of $250 for each fight. If the bulls are tame the show is fined for giving a poor performance and swindling the people. The *matador*, El Capitan, whose duty it is to strike the bull's heart with a sword, gets the highest salary, as much as $200 a performance; the other fighters receive from $10 to $100.

Sometimes a fight is given for charitable purposes. Young girls dressed like brides in white satin, veil and satin shoes, do all the directing, and young men of position and birth are the fighters.

It is to be supposed that when a man is killed in the ring the fight would stop, but that only seems to whet their desire for more blood, and a dead man is pulled off the field and another takes his place amid increased enthusiasm. At a fight two weeks ago one man was gored almost to death, another had his arm broken, and a woman, who had witnessed this from her seat, entered the ring and tried to kill the bull. She was caught on its horns and carried once around the ring and whirled around in her perilous position like a top. The audience shouted and was much disappointed when the bull cast the woman to the ground, devoid of clothing and badly bruised, but alive. At another fight three men were killed. Both times the spectators could hardly be forced to leave at the end of the performance. It is safe to assert that that beef sold at a high price.

Bernardo Javino, the man who was gored almost to death two weeks ago, has quite a history. He came from Spain fifty-one years ago, and is eighty-two years old, the oldest fighter in Mexico, and the most famous. He has fought in every bull ring in the republic, and has killed four thousand bulls. Senor Javino is a well-built, fine-looking fellow, and though but lacking eighteen years of one hundred is as strong as a man of thirty-five. He is a great favorite, and has received numerous and costly presents, among which he numbers one thousand fine bulls. But he is to-day very poor, and has only his salary. He is unmarried. Though the idol and favorite of the people, they shouted with joy when they saw him being gored. The bull caught him in the small of the back, and though making only one wound outside made five inside. He was carried off for dead, but though having a wound that would have finished any other man, he is still living, and asserts he will repay many bulls yet for his sufferings. The bull that had the honor to nearly finish the old warrior, killed three horses, broke the man's arm, and almost finished the woman.

Senor Javino has a nephew, Juan Moreno, who gives promise of being the best fighter, after his uncle, in the Republic. He is a six-footer of magnificent build, with a handsome face, fair complexion, with brown hair, resembling a handsome American boy, in honor of which the Mexicans have named him El Americano (the American). Their shouts are long and long for El Americano, and presents are showered down on him. He can accomplish the daring feat of striking the bull's heart with one thrust of the sword, which he withdraws instantly. This is considered scientific, for when the sword strikes the heart it is very difficult to withdraw, and is most always left sticking in until the bull dies. In the frontier the horns are sawed off the bulls before they go in the ring, in order to make the fight fierce and bloodier. It is said they are trying to stop this cruel torture.

The fight being finished the bands depart and the people make their way to the train with reluctance, where venders earn a mint of money by selling them pulque and a mixture of crushed corn and red pepper, done up in corn husks, which is eaten with a relish. After this Mexican feast is finished the train pulls out, everybody, men, women, and children, light their cigarettes, and between puffs they discuss the merits and demerits of the fight. The homeward trip is a very joyous one, so much so that "the daisy policemen" are often called on to exert their influence in quieting the mirth.

CHAPTER IX.
THE MUSEUM AND ITS CURIOSITIES.

THE first place tourists go on reaching Mexico is to the post-office. All one has to do when desiring to know what the latest incoming party looks like, is to take a position near the post-office. They stroll up the street, generally "goose fashion," stopping now and again to gaze at some prostrated pulque drinker; a wardrobe moving up the street on a pair of bare legs – *i. e.*, a woman with a half-dozen babies tied to her; an old cripple sitting on the walk selling taffy, or a blind man selling lottery tickets. Amid all this they manage at last to get into the office, and we see them emerge, a half-hour later, with funeral-like faces, and woman-like tongues giving their opinions of the officials who do not understand bad Spanish, not to mention English, and of the mails which take three days and the same number of nights to come from the nearest point of the States, El Paso.

For the want of something better to do we will follow them to the next point of interest – the museum – which is in the same building, several doors above the post-office. It is not the kind of a museum where you have a two-cent show for a ten-cent silver piece, but it is a place that any city might be proud of. At the top of the stairs, for the museum is on the second floor, are several large paintings of religious subjects and an immense mirror with a fine frame, which was stolen from some cathedral during one of the many revolutions of Mexico.

The first room contains a life-size portrait of Maximilian, seated on a beautiful white steed. Around are Mexicans gazing at him with admiration and awe. Maximilian is a handsome man, and the picture is said to be the finest of Maximilian in existence. If so, he was indeed, by virtue of looks, worthy to be an emperor.

In the center of the room on a table is the silver service, composed of one hundred and seventy-six pieces, used by Maximilian and Carlotta. Each piece bears the arms of the empire and the mark of the factory "Cristofle." It is massive and elegant; little silver cupids with wreaths of flowers are placed in every available spot. Many of the pieces are a load for two men. A bronze bust, life size, of Maximilian, has decorations and ten halberds, silver-mounted with blue and gold trimmings, ordered by the emperor to be used by the Palace Guard on state occasions, are all placed side by side. In a case in the same room are a number of loose pieces of armor worn by the conquerors. Two pieces, a breast plate and helmet, have the name of "Pedro de Alvarado," the Spanish captain who made the world-famous leap near Noche Triste.

Portraits of sixty-two Spanish Viceroys line the room. They were removed from the national palace here, on the establishment of the independence of the Republic. The frames are of black wood and the paintings are old style. It may have been the fashion in the day of white queues to always have one "off" eye, for one eye in nearly all the pictures goes a different direction from its mate, and in many instances the "off" eye is as roguish as a little brother, making you imagine the old rascals are going to wink, while the opposite orb gazes out in saint-like expression. The effect is ludicrous. The glass-ware of the Emperor Iturbide, containing excellent portraits of himself and Chapultepec Castle, is also shown in this room. In the next room, in a glass case, lying on a red satin, gold covered pillow, is a plaster paris cast of the face of Juarez, the much beloved Indian President; hairs of his head are still adhering to the plaster, and it is certainly the finest thing of the kind ever executed.

The portraits of Fernando Cortes Agustin de Iturbide, Emperor I., Ignacio Allende, one of the earliest patriots of Mexico, the great Antonio Lopez de Santa Ana and Don Vicente Guerrero, who was the third President of the Republic, are here, to say nothing of other things of historic value, such as the arms of the Mexican Republic made in 1829, surrounded with Indian mosaic feather work; an old, worn damask banner used by Cortes in his second expedition against the great Montezuma, and the arms of the city of Texcoco, presented by Charles V., of Germany, and Charles I., of Spain.

The little idols perhaps attract more attention than anything else at the museum. In two long rooms the cases lining the walls are filled with idols of all sizes and shapes, made of stone onyx and marble. Some of the pottery is horribly exquisite. Beads used by the Indians, made of stones, teeth and bones, are numerous. The large objects on the pedestals come in for a share of wonder. They are adorned with names of wondrous length and non-pronounceable, and stories of horror. Izcozauhqui (the Fire of the Sun) is in ugly red and yellow clay; Huitzilopoxtli (the God of War), a black clay image, equally ugly. A clay urn with carved faces, flowers and fruits on the outside, is called the "Funeral Urn." The "Goddess of Death" is an image some fifty inches in height, with large round eyes formed of bone, and outstretched hands of the same material, her skirts are formed of serpents and her head is a skull. Large brown earthen jars, said once to have held sacred fires, are among the collection. It may be historically correct and all the horrible tales connected with these things true, but the more one looks the less probable it seems, and after all they may have been innocent statues and flower vases used by this people in former days. It is just as likely, and easier to be believed, for how can it be asserted, when they are unearthed after centuries, that they were used for any special purpose. Of course, the more sensational the story the better for print, but it is much easier to believe they were only harmless objects in some park or flower garden.

History tells us the Aztecs knew no alphabet, and used in place certain signs or figures for every subject – history, religion, feasts, wars, famines, and even poetry. The art of writing in this manner was taught by the priests, and handed down from father to son. Painters had to be frequently called to decipher the documents, and were treated with the highest consideration by the nobility. The manuscript employed was made of maguey and other plants and of skins. The Spanish destroyed the majority of these manuscripts, which would have been of great value if preserved. A few are now in the museum. From an artistic point of view, they are horrid.

The colors they used in painting are nearly always indelible and very bright. One of the paintings shows a snow-capped mountain, Popocatepetl, and to the left the City of Mexico, entirely surrounded by water. A fifty-foot maguey paper painted in black, contains the history of the Aztecs. How they left an island which held a temple and came to Mexico, establishing the city, with all the principal events which befell them in their wanderings. The battle of Noche Triste and the advent of the Spanish, are carefully portrayed. This is one of the famous picture writings, which are too tiresome to enumerate further.

The feather shield which belonged to Montezuma II., is in a frame in the same room with the picture writing. It is an old, worn-out, faded thing, and hangs too far away to be seen well. It was among the curiosities given by Cortes to the Emperor Charles V, He in turn presented it to the Museum of Vienna, where it remained until Maximilian restored it to Mexico.

One room is devoted to the display of Mexican marbles, stones, ores, etc. Another has petrified snakes, wood, human and animal bones. Cow horns measuring seven feet from tip to tip were excavated somewhere near Mexico. Elephant jaws and tusks which treble the size of those sported by the late lamented Jumbo are also from the historic, mysterious earth of Mexico. Among the many other things were noticed human bones protruding through a rock, and a turtle's shell which, if opened, would make a carpet for a grand salon.

Snakes, lizards, fish, and crabs of all kinds fill one good sized room, divided in the center by stuffed alligators, sword fish, crocodiles, and boa constrictors. This opens into another department, and here you meet the Mexican dudes occasionally. There are few collections of birds to equal this. Added to their own numerous beautiful and rare birds are specimens from all parts of the world. The work is especially fine, and the birds and fowls appear as if in life. One thing to be regretted is they have no butterflies. In all the museum they have but one small case, and they are the beauties which come from Brazil. The collection of beetles is somewhat larger, but still is nothing remarkable.

Monstrosities are quite plenty. One little calf has one head, one large eye in the center of its forehead, and two perfect bodies. Another has one perfect body and two heads. Two pheasants are fastened together like the Siamese twins. Dogs, cats, chickens, and even babies come in for their share of doubling up into all kinds of queer shapes. Monkeys, baboons, gorillas and a dilapidated elephant and giraffe finish this interesting quarter.

The court of the museum is planted with beautiful flowers and trees. Large idols were once standing there, but they have been moved inside of the building opposite the entrance. The idols can lay no claim to beauty, and are anything but interesting, except to people who have a wonderful amount of faith and a capacity to believe a fellow-creature's wild imagination. Scientific gentlemen with long faces and one eye-glass gaze at the images and translate, or at least pretend to, the hieroglyphics which cover them. We would not think for a moment of putting an opinion against one held by wise men since the time of the Conquest, and we would not like to say Bernal Diaz had an object in making the Indians as black as possible, but we would like to gently hint our little observations.

The sacrificial stone, where they claim fifty thousand people have been sacrificed, looks little as if intended for that bloody purpose. The stone is perfectly round, between four and six feet across and about two feet in thickness. On the upper side is sculptured the image of the sun or moon and on the sides, are groups of men, fifteen in number, and fifteen separate groups. Certain hieroglyphics accompany each group. The work is fine, and must have been done with great care and patience by a master hand. Marring the top is a rudely cut hole with a shallow groove running to the edge. If these people were making a sacrificial stone would they have cut fine figures, requiring care and time, and then spoil them by cutting out a big hole? Would not the basin have been cut out finely and the carvings made to fit? I may be lacking in knowledge and faith, but I have tried to believe, have gazed on the stone with the thought, "History says the blood of fifty thousand human beings has dripped down over that stone," but proofs assert themselves, and the poor scandalized thing seems to hold up every side and the ugly marring of its beauty, and reply, "Now, do I look as if I was made for that purpose?"

Though believing it was nothing more than an innocent Aztec calendar, we will repeat the sensational legend that covers it with a bloody cloak. There existed an Aztec order which worshiped the sun, and on this stone, they sacrificed human beings, calling them the "messenger to the sun." The "messenger," who was always a prisoner, was painted half red and half white. Even his face was divided in this manner. A white plume was glued to his head. In one hand he carried a gaily trimmed walking-stick, and in the other a shield with cotton on it, and on his back was a small bundle of different articles. Music was played as he ascended the stairway to the temple. There he was greeted by some high priest, who commanded him to go to the sun, present the articles he carried and deliver messages they sent. Finally, when he reached the summit, he turned toward the sun and in a loud voice proclaimed what was told him. Then they took away his bundles and cut his throat, drenching the sun on the stone and filling the bowl with his blood. When the blood ceased to flow the heart was cut out and held aloft to the sun until cold. Then the message was delivered.

It is said the Aztec calendar was carved in 1479, and its inauguration was celebrated with fearful sacrifices, but the conquering Cortes had it pulled down, and it remained buried until lowering the grade of the ancient pavement in 1790, when it was built in the southwestern tower of the cathedral. There it remained until about a year ago, when it was removed to the museum, where it now occupies a prominent position. The Sad Indian, a statue so-called because it was unearthed on a street of that name, is a jolly-looking fellow, and compels one's admiration, despite his broad nose and ugly features. So far, I have heard no blood-curdling tales connected with him, but the wiseacre shakes his head solemnly and replies: "Hundreds of human beings were sacrificed on his account, but the history has escaped my memory." Meanwhile, the old fellow sits there with folded hands and a comical expression on his face, thinking, probably, of the duties which he once performed, which were, undoubtedly, holding a lamp or a flag, as the hole through the folded hands and between the feet directly beneath proves.

It is quite interesting to roam around and examine this broad face and that slim one, from those of mammoth size to ones the size of one's hand. We grow to like the queer objects which certainly formed some part in the lives of those strange people who lived and died centuries before us.

In one corner locked up in a cell by itself is the coach of Maximilian and Carlotta. It is one of the finest in the world, and is similar in construction and finish to that used on State occasions by the Czar of Russia. The coach was a present to Carlotta from Napoleon II. It is so fine that it is difficult to give a description of it. The royal coat of arms is on every available spot, on the doors and above, wrought in gold, and embroidered in gold on the crimson velvet which covers the driver's seat. The entire coach is gold and crimson except for the inside, which is heavy white silk, cords, fringe and tassels of the same. Gold cherubs the size of a three-month-old baby finish each corner. The carriage was drawn by eight pure white horses or the same number of coal black ones, and as it swept down the grand passes to superb Chapultepec, holding its royal owners, it must have been a sight fit for kings. But it stands to-day a silent memento of a murdered young emperor and a blighted empress. All the men employed at the museum are disabled soldiers, and it speaks well of the government to give them this employment. They seem to rightly belong in among this queer stuff, for it would take half a dozen of them to make a whole man. The museum is open only from ten to twelve, and is free to all.

But our tourists are even now standing on the outside, wondering if they have not fasted enough to do penance for all the sins ever committed; and if they will get much else than frijoles, rice, and red peppers for dinner – or, more properly speaking, breakfast. We know just what they will visit this afternoon, and if you care to see it also we will try, in our humble way, to show you around.

CHAPTER X.
HISTORIC TOMBS AND LONELY GRAVES.

HOW much I would like to paint the beauties of Mexico in colors so faithful that the people in the States could see what they are losing by not coming here. How I would like to show you the green valley where the heat of summer and blast of winter never dare approach; where every foot of ground recalls wonderful historical events, extinct races of men and animals, and civilization older by far than the pyramids. Then would I take you from the table-laud to the mountain, where we descend into deep canons that compare in their strange beauty with any in the world; the queer separation of the earth, not more than 100 feet from edge to edge of precipice, but 400 feet deep. More wonderful still is the sight when the rainy season fills these gorges with a mad, roaring torrent. Then would I lead you to the edge of some bluff that outrivals the Palisades – and let you look down the dizzy heights 500 feet to the green meadows, the blooming orchards, the acres of pulque plant, the little homes that nestle at the foot of this strange wall. Then further up into the mountains you could see glaciers, grander, it is claimed, than any found in the Alps. Here are buried cities older than Pompeii, sculptures thousands of years old, hieroglyphics for the wise to study, and everywhere the picturesque people in their garb and manners of centuries ago – and all this within a day's travel from the city. Surely in all the world there is none other such wonderful natural museum.

Business men who wish to rest from their labors find perfect quiet in this paradise. All cares vanish. Some strange magic seems to rob one of all care, of every desire to hurry. Railways furnish comfortable and safe transportation; the people are attentive and polite, and as many comforts are attainable as at any other place away from the States. People who have any desire to see Mexico in all its splendor should come soon, for civilization's curse or blessing, whichever it may be, has surely set a firm foot here, and in a few years, yielding to its influence, all will be changed. Already the dark-eyed senora has changed the lovely, graceful mantilla for stiff, ugly bonnets and hats; the poor Indian woman is replacing the fascinating reboza with a horrid shawl; the Indian man is changing sandals for torturing shoes and the cool linen pantaloons and serape for American pantaloons and coat. Civilization and its twin sister, style, have caught them in their grasp, and unless you come soon Mexico will cease to be attractive except as a new California.

There is one thing I hope will ever remain, and that is the graveyard of San Fernando, where most of the illustrious dead of Mexico are entombed. But it is doubtful, as a little beyond are the fine houses of the foreign representatives, and the houses are crowding up to the gate of this dead city as though trying to push it out of existence. An old cathedral, faced by a green plaza, rears its head at one side, near the massive iron gates which the keeper, sitting just within its portals, swings open and admits one with a welcome that is surprising. All around are people buried in the walls. The plates are decorated in all manner of ways. Some have a little niche which hold the image of the Virgin and several candles. Others are hung with wreaths, and some with crepe. The majority have places to hold candles, which are burnt there on certain days. The nearest tomb to the gate holds the remains of a young girl who died, quite suddenly, on the day she was to be married, just an hour before the time appointed. Near here is erected a fine shaft in honor of General Ignacio Comonfort, who was a President once, but was shot at Molino de Toria, November 13, 1863, by the Americans. Several yards beyond is a plain, brown stone, built in an oblong box shape, with a large, stone cross in the center. It is weather-beaten and worn, and looks to be centuries old. All the information it gives a stranger is in two large initials, T. M., rudely cut on the side.

No date or usual verse of regret from loving friends is inscribed, and somehow a thrill of pity strikes one for T. M., as it seems to be the only grave in all that quiet city that bears no mark of loving hands. I took my penknife and hastily cut in the soft stone R. I. P. "When the Mexican friend, who had during this time been engaged with the gateman getting some information, came up he said: "The grave you stand beside is that of General Tomas Mejia, who was shot with Maximilian, and here is the tomb of the other." It was similar in shape to General Mejia's, but some kind hand had hung wreaths on the cross. General Miguel Miramon was president of the republic before Maximilian. He was a brave and good man, and the emperor well knew his worth.

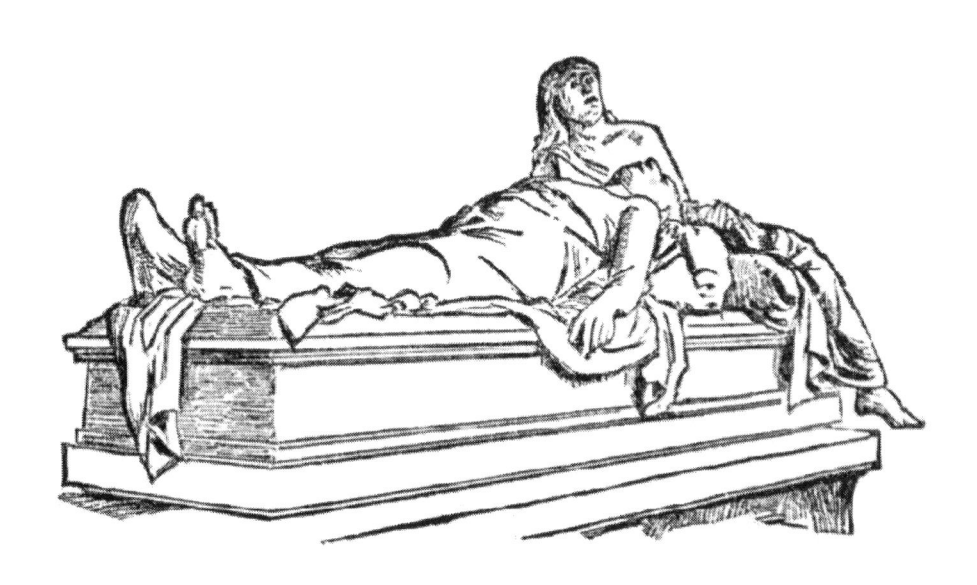

TOMB OF BENITO JUAREZ.

When they stood up to be shot, Maximilian in the center, Mejia on the right and Miramon on the left, the center of course being considered the place of honor, Maximilian, touching Miramon on the shoulder, said: "You are more worthy this place than I," and he exchanged places, and so they died.

The tomb of Benito Juarez, the Indian President, is the finest in the place. It is a long marble tomb. On it lies the life-size body of Juarez, partly covered with a mantle. Sitting at his head, with her hands on his heart, is a beautiful woman, representative of the nation mourning for its much beloved President. The whole is a perfect study, and was designed and executed by a Mexican.

The life of Juarez is a very romantic one. He is familiarly known as the "Lincoln of Mexico." He was born in the State of Oaxaca, 1806, and at the age of twelve years could neither read nor write. He was a full-blooded Indian, and could not even speak the Spanish language. However, he tried to improve his time, and in 1847 he was Governor of his native State. He went to New Orleans, on being banished by Santa Anna, but returned to Mexico in 1855 and became President of the Court of Justice. When Comonfort was overthrown by the clerical party, Juarez set himself up at Vera Cruz as Constitutional President of the Republic. The United States recognized him as such, and he successfully fought the priesthood and confiscated all the church property. When Maximilian ascended the throne, Juarez sent his family to New Orleans, but he remained here until compelled to cross the frontier. The United States, which had always favored Juarez, interfered in his behalf. At the termination of the War of the Rebellion Maximilian was betrayed and shot, and Juarez was re-elected in 1871, and died in office June 18, 1872.

He has a daughter who is married and living in Mexico in greater style than the president. She resembles her father. A story is told of Juarez that is new at the very least. He had plenty of enemies, especially among the church party. One day he sent a band out to capture an outlaw, who, notwithstanding his enemies, stood well with the clergy. The bandit was met on the highway and shot before he could utter a prayer. They said his soul was lost, and Juarez was to blame. When he was dying it was endeavored to keep the matter quiet, and the people were in ignorance of his fatal illness until one morning they saw a notice posted on street corners, which read in this style:

"Hell, 1.30. Juarez just arrived. Devil putting on his tail."

It was signed by the name of the bandit.

General Ignacio Zaragoza, the conqueror of the French in Puebla, May 5, 1862; General Vincente Guerrero, one of the principal heroes of the War of Independence; Mariano Otero, one of Mexico's most famous orators; Melchor Ocampo, a very distinguished philosopher and politican, and the companion and right hand of Juarez, helping him to establish the liberal principles; Francisco Zarco, one of the Constitutionalists; General Jose Joaquin de Herrera, one of the best Presidents the Republic ever had, and other famous generals, statesmen, writers, and artists fill up this quiet spot. The gates are only open now to visitors. They no longer register dead guests.

Among many other things Mexico can boast of is the public library. It is situated on Calle de San Augustin, in the old church and convent of Saint Augustin. The high iron fence which incloses it is topped with marble busts of famous orators and authors. The little green plot in front is filled with rare plants and fountains. The face of the church is a mass of wondrous carvings, and the vestibule is a crown of splendid architecture. Directly over the door leading into the room is the "World." On one side brass hands and figures tell the hour. Standing on one foot on top of it is a life size figure of "Time," in bronze. The attitude, the scythe over the shoulder, the expression on the face, the long, flowing beard and hair are perfect. Opposite Time, and at the other end of the room is the Mexican coat of arms. Book-cases line each side, and in the center, are reading-desks and easy-chairs. At the right entrance is a large statue of Humboldt, and on the left Cuvier. Opposite one another are Descartes and Copernicus, Dante and Alarcon, Origen and Virgil, Plato and Cicero, Homer and Confucius, and in the center a large figure with a book in hand marked "Science."

The books are catalogued under the heads of philosophy, history, fiction, etc., and are placed in cases alphabetically. They are in all languages, and many of them are very ancient. Some are on parchment and in picture writing. The library has catalogued one hundred and sixty thousand volumes, and owns many besides that are not yet sorted and arranged. It is open from 10 A. M. to 5 P. M., and is equally free to all. It is well patronized by men, but it is safe to say no woman has ever read a book inside its walls. The only women who ever enter are tourists. The books are not permitted to go outside the building. A man gets a printed card. On it he writes the title, number and case of his book, and when the hour comes to close he lays the book on the desk of the janitor and gives his card to the superintendent. Many of the ancient books were taken at the time of the confiscation of the monasteries and convents.

The carnival passed off very quietly. As I said before, Mexico is becoming civilized, and doing away with many ancient and beautiful customs. In former years every day on carnival week the paseo was crowded with masked men and women in historic and comic garb, and battles were fought with empty egg-shells and queerly constructed things for the same purpose. This year every person went, but only the fewest number were masked. Some few among the lower class threw egg-shells. Beyond this all was quiet. It has also been the custom to give fancy-dress and masked balls. In all the theaters public balls were held and the clubs gave private receptions. The French Club had their rooms nicely decorated and the best people attended, dressed in the finest and most original costumes. Perhaps the most striking one was a creamy satin embroidered with red roses and covered with natural butterflies of gorgeous and brilliant hue. The young ladies all wore their dresses just reaching their knees, and the fancy boots displayed were something marvelous; satin of all shades, embroidered with gold and silver, and trimmed with flowers.

One couple, who have been lately engaged, were dressed alike. The girl wore a short dress of white satin, profusely trimmed with pompons of white fur; white satin boots trimmed the same way, and over her loose hair of marvelous length and thickness was a point lace veil. The groom wore satin knee breeches, short coat, high hat and boots, all covered with the white fur pompons. They were accompanied by the mother, in a brocade crimson velvet on a canary background and rich yellow lace, low-necked and en train, and the father in common dress suit. The Mexican boys never appeared better than in the grand old dress of former days. Mostly crimson velvet and satin were affected, showing to an advantage their superb eyes and complexion. The women were remarkable for their homeliness.

A grand supper of thirty-five courses was served and more wine, champagne and cigarettes consumed than would be done at forty receptions in the East.

Now, having shown you how they do at private balls where only the *elite* are permitted to attend, would you like to don a mask and domino and sit with these very same people in the boxes at the theaters, and watch the promiscuous crowd beneath? It is not a select crowd by any means, but one composed of the lowest in the land. Yet men take their wives, sisters, and friends, masked, that they may watch through opera glasses this wonderful sight, and wives and sweethearts get friends to take them, that they, unseen, may see if husband or lover takes part in the revel, for the men are of the best and wealthiest families.

At 11 o'clock the doors are flung open and people come in slowly. The two bands play alternately the Spanish danza and the waltz. The women come in dressed in all the styles ever invented. One beautiful woman wore a blue satin dress, embroidered with pink rose buds. Another wore blue, trimmed with beaded lace, which glittered like hundreds of diamonds in the gas-light. Two came together, one in black, the other in crimson velvet, profusely and gayly embroidered. Some were dressed after the style of the male dudes of the States, but the majority wore nothing but a comic-opera outfit, dotted with silver or gold spangles, according to the color. The men, with the exception of a half dozen, wore their common suits, and never removed their hats. Nearly all the women wore their hair short, which they had powdered.

At first, they wore masks, but in a short time they were removed, and by 3 o'clock everybody was drunk. When a man refused to dance with a woman, a fight was the result, and everybody would quit dancing until it was settled. One year fifteen men were killed during the week it lasted. This year but one has met his death. The actions and dancing of this mob will bear no description, and at 7 o'clock the performance ended. The manager of the National Theater has promised that his house shall never be used for this purpose again.

The carnival was celebrated in fine style at Amecameca, right at the foot of the White Lady. Indians came from all parts of the country and paraded the entire week around the church and temple with lighted candles. At Puebla they had egg battles, and in all little places the feast was carried on as in former days.

Sights in the city have begun to assume a familiar look, although one never tires of them, and I begin to think of moving elsewhere.

The buried city is slowly being unearthed at San Juan. Already they have brought to light a house of magnificent size and finish, and in a few days, it will be well worth a visit. Tourists have been going down regularly, but beyond a few men at work, little was to be seen. What they missed they furnished with their imagination, as did also some correspondents who would not wait to get legitimate news.

The mint, which is situated in the suburbs of the city, is turning out fifty thousand dollars in silver per day. The first coin struck was in 1535, and in three hundred years they coined $2,200,000,000. The men employed get from one to two dollars a day. In a month from now the government is going to make fifteen million cents. Gold coin, although in use here, is not made more than once a month.

The arsenal is in a fine old building directly in the opposite direction from the mint. All departments are not running – for the lack of money, so they say. They make but three hundred and fifty entire guns a day, but have one million dollars' worth in stock. In one room they have a fine collection of arms, such as are used by every nation in the world. The iron and wood used is Mexican, the latter a superb walnut, which requires no oil or varnish. The people here employed get from one real (twelve and a half cents) to two dollars a day, the highest that is paid.

The tourists who have such a mania for mementos have brought disgrace on themselves and others also. The governor has been very kind, and has thrown open the ambassadors' hall, without reserve, for their inspection. It is a beautiful place, containing life-size paintings of Washington, Juarez,

Hidalgo and other illustrious men. The chandeliers, hung with brilliant cut-glass pendants, terra cotta and alabaster vases and handsome clocks, were once the property of Maximilian. At either end of the long hall are crimson velvet and gold-hung thrones, where the president receives his guests. Some trophy fiend, most probably some girl with the thought of a crazy patch, cut a large piece out of one of these damask curtains; consequently, the governor has issued orders that no visitors shall be admitted, and the Yankees have gone down one notch further in the scale where they already, by their own conduct, hold a low position. It is to be hoped that those who come in the future may act so that no more shame will fall on us.

CHAPTER XI.
CUPID'S WORK IN SUNNYLAND.

LOVE! That wonderful something – the source of bliss, the cause of maddened anguish! Love and marriage form the basis of every plot, play, comedy, tragedy, story, and, let it be whispered, swell the lawyer's purse with breach of promise and divorce case fees. Yet it blooms, with a new-found beauty in every clime, and as there is no land in all the world more suitable for romance than Mexico, it is pertinent to show how love is planted, cultivated and reaped in this paradise, so as to let our single readers in the States compare the system here with home customs and benefit thereby, whether by making good use of their own free style or cultivating a new, those interested must decide.

Mexicans may be slow in many things, but not slow in love. The laws of Mexico claim girls at twelve, and boys at thirteen years are eligible to marriage, and it is not an unusual sight to see a woman, who looks no more than thirty-five, a great-grandmother. As children, the Mexicans are rather pretty; but when a girl passes twenty she gets "mucho-mucho" avoirdupois, and at thirty she sports a mustache and "galways" that would cause young bachelors in the States to turn green with envy. The men, on the contrary, are slim and wiry, and do not boast of their hirsute charms, especially when in company with women, as they have little desire to call attention to the contrast, and the diamond-ring finds other means of display than stroking and twisting an imaginary mustache. Yet this exchange of charms interferes in no way with love-making, and the young man wafts sweet kisses from his finger-tips to the fair – no, dark – damsel, and enjoys it as much as if that black, silky down on her lip were fringing the gateway to his stomach.

Boys and girls, even in babyhood, are not permitted to be together. Before very long they compel their eyes to speak the love their lips dare not tell, and with a little practice it is surprising how much they can say, and how cold and insipid sound words of the same meaning in comparison.

All the courting is done on the street. When evening kindly lends its sheltering cloak, even though the moon smiles full-faced at the many love-scenes she is witnessing, the girl opens her casement window and, with guitar in her hand or dreamily watching the stars, she awaits her lover. If her room is on the ground floor she is in paradise, for then they can converse – he can even touch her hand through the bars. But if she is consigned to a room above she steps out on the balcony. If the distance is not too great, they can still converse; but otherwise, with the aid of pencil, paper, and tiny cord, they manage to spend the evening blissfully without burning papa's coal and gas, and staying up until unseemly hours.

The lovers are unmindful of the people who pass and repass, and the kind-hearted policeman never even thinks of telling the young man to "move on." If the house is secluded the lover tells his devotion in musical strains. Night is not only devoted to love-making, but in the broad light of day the young man will stand across the street and from the partly opened casement of the fair one, is visible a hand and a nose – of course she has full view, but that is all that can be seen of her. With the hand they converse in deaf and dumb language, which, added to their own signs, makes a large dictionary. It is not likely there exists a Mexican who is not an adept in the sign language. Courting is too vulgar a word for them, so they call it – translated in English – playing the bear.

You would naturally wonder how a girl who never leaves her mother's or chaperon's side, who never goes to parties, who is watched like a condemned murderess, would ever get a lover; but notwithstanding all this strictness they number less old maids and more admirers than their sisters in the States.

Perhaps while out driving, at the theater or bull-fights, they see a man they think they will like. He is similarly impressed. He follows his new-found one home, and she knows enough to be on the balcony awaiting his arrival with the shades of night. He may play the bear with her for a year and she not know his name. He has the advantage, for he can find out everything about her family, and thereby determine whether she is a desirable bride or not.

Sometimes they play the bear for from seven to fifteen years – that is, if the parents are very wealthy – and even then, not get the girl, for with all their passionate love they number many flirts. Often one girl will have two or more playing the bear at the same time. If they chance to meet they inquire, fiercely, "Whom are you after?" If the answer demonstrates the same girl, one will request the other to step aside. if he refuses a duel follows. After that, the girl is bound, by the custom of her country, to relinquish both. If a brother or father discovers a "bear," the latter must submit to a thrashing from their hands if he still desires to retain the girl's love. If a father notices the attention of a "bear" and looks with favor upon him, he does not disturb his "playing." When he concludes he has served long enough he is invited into the house. This means the same as if he had asked her hand in marriage and has been accepted. He is the intended husband, but never for a moment is he alone with his *fiancee*. He may aspire to take the driver's place sometimes, or to take the entire family to the theater.

A young American had been received in great favor by a Spanish family; probably the old man thought he would like an American for a son-in-law. However, young America was not going to waste any time sitting in the house with the old folks, so he politely requested the object of his admiration to go to the Italian opera. She graciously accepted. When he went to the house he found not only his lady love but the entire family prepared to accompany him. The deed was done; he could not back out, and for the privilege of talking to the mother, with the daughter sitting on the other side casting love-lit glances from her splendid eyes, he paid forty-three dollars. He was disgusted, and accordingly gave up his chance of being a member of a Mexican family.

If a man gets impatient and feels like becoming responsible for the price of his sweetheart's bonnets, he asks the father. If he is rejected he can go to a public official, swear out a notice to the effect that his and the girl's happiness is ruined by the father's heartlessness. He then secures a warrant, which gives him the privilege of taking the girl away bodily from the home of her parents. This is a Mexican elopement. If, on the other hand, he is accepted, the wedding-day is named, and agreements are drawn up as to how much will be the daughter's portion at the death of her parents. Before that period, she receives nothing. The intended husband furnishes the wedding outfit, and all the wearing apparel she has been using is returned to her parents. She has absolutely nothing. The groom buys the customary outfit – white satin boots, white dress and veil.

A Mexican wedding is different from any other in the world. First a civil marriage is performed by a public official. This by law makes the children of that couple legitimate and lawful inheritors of their parents' property. This is recorded, and in a few days – the day following or a month after, just as desired – the marriage is consummated in the church. Before this ceremony the bride and groom are no more allowed alone together than when playing the bear. At a wedding the other day the church was decorated with five hundred dollars' worth of white roses. The amount can be estimated when it is stated roses cost but four reals (fifty cents) per thousand. Their delicate perfume filled the grand, gloomy old edifice, which was lighted by thousands of large and small wax candles. Carpet was laid from the gate into the church, and when the bridal party marched in, the pipe organ and band burst forth in one joyous strain. The priest, clad entirely in white vestments, advanced to the door to meet them, followed by two men in black robes carrying different articles, a small boy in red skirt and lace overdress carrying a long pole topped off with a cross.

The bride was clad in white silk, trimmed with beaded lace, with train about four yards long, dark hair and waist dressed with orange blossoms. Over this, falling down to her feet in front and reaching the end of the train back, was a point lace veil. Magnificent diamonds were the ornaments, and in the gloved hands was a pearl-bound prayer-book. She entered a pew near the door with her mother – who was dressed in black lace – on one side and her father on the other. After answering some questions, they stepped out, and the groom stood beside the bride, with groomsman and bridesmaid on either side, the latter dressed in dark green velvet, lace, and bonnet. The priest read a long while, and then, addressing the girl first, asked her many questions, to which she replied, "Si, senor." Then he questioned the groom likewise. Afterward he handed the groom a diamond ring, which the latter placed on the little finger of the left hand of the bride. The priest put a similar ring on the ring-finger of the right hand of the groom, and a plain wedding ring on the ring-finger of the bride's right hand. Then, folding the two ringed hands together, he sprinkled them with holy water and crossed them repeatedly. The band played "Yankee Doodle," and the bride, holding on to an embroidered band on the priest's arm, the groom doing likewise on the other side, they proceeded up to the altar, where they knelt down. The priest blessed them, sprinkled them with holy water, and said mass for them, the band playing the variations of "Yankee Doodle." A man in black robes put a lace scarf over the head of the bride and around the shoulders of the groom; over this again he placed a silver chain, symbolic of the fact that they were bound together forever – nothing could separate them.

After the priest finished mass he blessed and sprinkled them once more. Then from a plate he took seventeen gold dollars the groom had furnished and emptied them into his hands. The groom in turn emptied them into the hand of his bride, and she gave them to the priest as a gift to the Church and a token that they will always sustain, protect, and uphold it. Now the ceremony, which always lasts two to four hours, is ended, and the newly married pair go into an adjoining room to receive the congratulations of their friends. The marriage festivities are often kept up for a week. After that the husband claims his bride, and right jealously does he guard her. Her life is spent in seclusion – eating, drinking, sleeping, smoking. The husband is desperately jealous and the wife is never allowed to be in the company of another man. Life to a Mexican lady in an American's view is not worth living.

When death takes one away the dust remains buried for ten years, if the husband is wealthy. At the end of that time the bones, all that remains in this country, are lifted, placed in a jar and taken home and the tombstone used as an ornament. "See that case?" said a Mexican. "My first wife is in that, even to her fingernails, and that is her grave-stone." So it was, there in the parlor, a dismal ornament and memento.

Mexican carelessness does not extend to the saying of mass. A man had three daughters, and each was to inherit $3,000,000. For this reason, he would not allow them to marry. One died, and the anniversary of her death was celebrated in fine style. High mass was said, and a coffin arranged on a catafalque forty-four feet high recalled the dead woman. The coffin, etc., were imported from Paris, and altogether the mass cost $30,000. That's dying in high style.

Mexicans who have been to the States much prefer the American style of calling on ladies, but it is not likely it will ever be the custom – for American residents here have adopted the Mexican style for their daughters, and most ridiculous and affected does it appear. American boys, however, have no time to waste on such manners, so they do their love-making by letters and go back to the States for their brides, leaving the American mammas to search among the Mexicans for ones to play the "bear."

CHAPTER XII.
JOAQUIN MILLER AND COFFIN STREET.

DEAR old Mexico shows her slippered foot, for summer is here. The fruit-trees are in blossom, the roses in bloom, the birds are plenty and everybody is wearing the widest sombrero. From 10 o'clock until 2 the sun is intensely hot, but all one has to do is to slip into the shade and the air is as cool as an unpaid boarding-house-keeper and fresh as a "greengo" on his first visit to the city. At night blankets are comfortable. Tourists are still flocking to Mexico, many with business intentions, and the United States at present is as well represented as any other foreign country. Yankees are looked on favorably by some of the better and more educated class of Mexicans, but others still retain their old prejudices. However, one can hardly blame them, for, barring a few, the American colony is composed of what is not considered the better class of people at home. They have come down here, got positions away above their standing, and consequently feel their importance; they are more than offensive, they are insulting in their actions and language toward the natives, and endeavor to run things. The natives offer no objections to others coming here and making fortunes in their land, but they have lived their own free and easy life and they do not propose to change it, any more than we would change if a small body of Mexicans would settle in our country; and we would quickly annihilate them if they would offer us the indignities the Americans subject them to here.

I dread the return and reports of such people in the States, for although there are good and bad here, the Mexicans have never been represented correctly. Before leaving home, I was repeatedly advised that a woman was not safe on the streets of Mexico; that thieves and murderers awaited one at every corner, and all the horrors that could be invented were poured into my timid ear. There are murders committed here, but not half so frequently as in any American city. Some stealing is done, but it is petty work; there are no wholesale robberies like those so often perpetrated at home. The people are courteous, but of course their courtesy differs from ours, and the women – I am sorry to say it – are safer here than on our streets, where it is supposed everybody has the advantage of education and civilization. If one goes near the habitation of the poor in the suburbs, they come out and greet you like a long absent friend. They extend invitations to make their abode your home, and offer the best they own. Those in the city, while always polite and kind, have grown more worldly wise and careful.

The people who give the natives the worst name are those who treat them the meanest. I have heard men who received some kindness address the donor as thief, scoundrel, and many times worse. I have heard American women address their faithful servants as beasts and fools. One woman, who has a man-nurse so faithful that he would sacrifice his life any moment for his little charge, addressed him in my presence as: "You dirty brute, where did you stay so long?" They are very quick to appreciate a kindness and are sensitive to an insult.

Speaking of honesty, they say the aquadores, or water-carriers, are the most honest fellows in the city. They have a company, and if anyone is even suspected of stealing he is prohibited from selling any more water. At intervals all over the city are large basins and fountains where they get their water. For four jars, two journeys, as they carry two jars at once, they receive six and a quarter-cents, or one real; twelve and a half cents if they carry it up-stairs. Their dress is very different from others. They wear pantaloons and shirt like an American and a large leather smock, which not only saves them from being wet but prevents the jars from bruising the flesh. They all wear caps, and the leather band of the jars is as often suspended from the head as from the shoulders.

Americans who come to Mexico to reside should take out identification papers the first thing. It costs but little and saves often a lot of trouble. People when arrested have little chance to do much even if they be innocent; they are thrown into prison and allowed to remain there, without a trial, for often a year, and it is said a Mexican prison gains nothing in comparison with Libby prison of war fame. But if a man has his identification papers he can present them and command an immediate trial, and it is given. There is an American now lying in prison here for shooting a Mexican woman; the woman was only shot through the arm, and yet the man has been in jail, without even a change of clothing, for over a year. He is in a deplorable state, without much hope of it being bettered. The American Consul seems to have a disposition to help his countryman. He has been here but a month, and his first work deserves praise. A man by the name of John Rivers, or Rodgers, shot a fellow in self-defense.

It was a clear case, but the main witnesses had no desire to lay in jail, as the law requires, until the American's trial came up, so they fled the country. The American could speak no Spanish. His trial was poorly conducted, and he was sentenced to be executed at Zocatagus, up the Central road. Consul Porch heard of the case. He studied it out, found the man was not given a fair trial, and hastened off, reaching the scene of execution but a short time before the hour appointed, but in time at least to postpone the tragedy. There is one great disadvantage Americans suffer from, and that is the government sending out ministers and consuls who have no knowledge of the language in the country to which they go. It would be a mark of intelligence if they would make a law, like that in some countries, providing that no man could represent America unless he had a complete knowledge of the foreign tongue with which he would have to deal.

In my wanderings around the city I found a street on which there are no business houses or even pulque shops – nothing but coffin manufacturers. From one end of the street to the other you see in every door men and boys making and painting all kinds and sizes of coffins. The dwelling houses are old and dilapidated, and the street narrow and dingy. Here the men work day after day, and never whistle, talk, or sing, as they go at their hewing, painting and glueing, with long faces, as if they were driving nails into their own coffins.

I soon related my discovery to Joaquin Miller, and he went along to see it. Then he said, "Little Nell, you are a second Columbus. You have discovered a street that has no like in the world, and I have been over the world twice. It's quite fine, isn't it?" and he gave a hearty laugh. Of course, there may be other streets somewhere just the same. We could find no name for our new treasure, so we simply dubbed it "Coffin Street." I am sorry I have no picture of it to send you, so you could see the coffins piled up to the ceiling; a little table in the center where the workman puts on the finishing touches, after which they are placed in rows against the building, by the sad-visaged and silent workers, to await a purchaser. Near this somber thoroughfare is another street where every other door is a shoe shop, the one between being a drinking-house. Many of the shoemakers have their shops on the pavement, with a straw mat fastened on a pole to keep off the sun. Here he sits making new shoes and mending old ones until the sun goes down, when he lowers the pole, and taking off the straw mat, furnishes a bed for himself in some corner during the night.

Wealthy Americans who have a desire to invest in land should come to Mexico. There is surely no other place in the world where one could get so much out of a piece of property. One end of a field can be tilled while the other is being harvested, and one can have as many crops a year as he has energy and time to plant. There is no doubt that anything can be cultivated here. Of course, peaches and apples are not plenty, because they only grow wild. Why, even a nurseryman would fail to recognize them in the small, scraggy, untrimmed bushes. The native fruits are fine, from the reason that they need no cultivating or trimming. If they did, Mexico would have a famine in the fruit line.

Land in Mexico is very cheap, and the Government collects a tax only on what is cultivated. One sensible man, by the name of Hale, came here from San Francisco a few weeks ago to buy property. A minister of the Gospel, a particular friend of Hale's, is authority for it that Senor Hale bought from the Government sixty-five thousand square miles – larger than the whole of England, I believe – for $1,000,000.

I don't think one would ever tire of the gayly-colored pictures Mexico is ever presenting. Though in Mexico two months, I can find something new every time I glance at the queer people. This little basket vender is but one of thousands, but we find he is the first one to wear his white shirt without tying the two sides together in a knot in front. He must surely have forgotten that part of his toilet, as it is the universal style and custom among them all. Very few Mexicans, even among the better class, wear suspenders. They wrap themselves about the waist with a bright-colored scarf, with fringed ends, and this constitutes suspenders. Many of the better class wear embroidered and ruffled shirt fronts.

The fruit venders have beautiful voices, and sing out their wares in such a charming manner that one is sorry when they disappear around the corner. They are sometimes quite picturesque with the fruit and vegetables tied up in their rebozo and baskets in their hands. Why the women have all their skirts plain behind and pleated in front I cannot say, but such is invariably the case. The men have horrible voices when they are out selling. There never was anything to equal them. I wonder if our florists would not like to buy orchids from the man who passes our door every morning with about a hundred of them strung to a pole which is suspended from his shoulder, only two reals (twenty-five cents) for exquisite plants, with the rare ones but little higher.

Mr. A. Sborigi, a Pittsburger, was in Mexico on a visit. When he landed in Vera Cruz he went into the country to see the place. Hearing music in a small cabin he drew nearer and recognized familiar tunes. "Wait till the clouds roll by," and Fritz's lullaby. A man came out and invited him in, and after a short time he said he was a colored man, that his name was Jones, and he came from Pittsburg, Pa. He is married to an Indian woman and has about twenty children, ranging all sizes. Mr. Jones is king of the villa. In one room he has a floor, a thing not possessed by any other inhabitant there, and his cabin is superior to all others. He is very proud of his wife and children, and has not the least desire to return to the Smoky City. He speaks Spanish, French, and English fluently.

When Mr. Sborigi was asked for his ticket on the Vera Cruz line, he jokingly handed the conductor an envelope that he had put in his pocket at New Orleans. On it was printed in English, "Tickets to all points of the world." The conductor took the envelope, looked at it, punched it and returned it to the donor. Quite amused, Mr. Sborigi tried it on others, and he not only traveled the entire distance to Mexico, but traveled on at least half a dozen branch roads leading from the Vera Cruz line to beautiful towns in the country. He took the punched envelope back to Pittsburg as a memento of the cheapest journey he ever took.

CHAPTER XIII.
IN MEXICAN THEATERS.

MEXICO does not know how a nation mourned for one Virginius like McCullough; has never witnessed Barrett's Cassius and David Garrick, or been thrilled with O'Neill's Monte Cristo; has never looked on Mary Anderson's exquisite form and cold, unsympathetic acting; has missed Margaret Mather's insipid simper and Kate Castleton's fascinating wickedness; is wholly unconscious of Little Lotta's wondrous kick and Minnie Palmer's broadness; has never seen pretty Minnie Maddern's "In Spite of All," and a mother of fifty odd years successfully transformed into a child of nine – Fanchon; is in blissful ignorance of "Pinafore" and "Mikado," and yet she lives and has theaters.

The most fashionable theater in Mexico is the National. President Diaz always attends, and of course the elite follow suit. It is well to say the president always attends, for there is little else to go to. Bull-fights, theaters, and driving are all the pleasures of Mexican life; the president gives no receptions or dinners, and entertains no Thursday or Saturday afternoon callers, so before death entered his family circle he was at the theater almost every night.

No paid advertising is done by theaters in the papers. Once in a while they, with the exception of the National, send around bills of their coming plays, accompanied by two tickets. For this they get a week's advertising; cheap rates, eh? Besides this they have native artists who select the most horrible scene to depict in water colors on cloth and hang at the entrance; these "cartels" are changed necessarily with every play, as billboards are in the States, and some of them are most ludicrous and horrible in the extreme. The Saturday I reached Mexico one of the theaters had on its boards a play, the cartel of which represented the crucifixion. What the play was could not be ascertained.

Sunday is the most fashionable theater day. Every person who can possibly collect together enough money goes, from the poor, naked peon to the Spanish millionaire. On Monday all amusement houses are closed and many are only open every other day throughout the entire week; they are not at all particular about fulfilling engagements. A play may be billed for a certain night and on arrival there the servant will politely inform you it is postponed until manana (to-morrow), and all you can do is to go back home and await their pleasure.

The National Theater is a fine building with accommodations for 4,500 persons. The first entrance is a wide-open space faced with mammoth pillars. Going up the steps you enter, through a heavily draped doorway, the vestibule or hall. Along the sides are racks where gentlemen and ladies deposit their wraps. The orchestra, or pit – the fashionable quarter in American theaters – is known as the "Lunetas." The seats are straight-backed, leather-covered chairs of ancient shape and most uncomfortable style. They were evidently fashioned more for durability than beauty, being made of very heavy, unpainted wood. Narrow passageways intersect each other, and wooden benches are placed along the seats to serve as foot-rests. Down in front of the stage is the orchestra, flanked at either end by long benches running length wise of the stage. Boxes, six stories in height, look out upon the stage, and balconies circle the room. The balconies are divided into compartments holding eight persons. Common, straight chairs, with large mirrors on the door and walls, are the only furnishment. The "Lunetas " command seventy-five cents to $1.50; Palcos (boxes) $2 a chair, and the Galeria (the sixth row of balconies) twenty-five cents.

At 8.30 the orchestra strikes up, people come in and find their places, and about 9 o'clock the curtain goes up and silence reigns; the enthusiasm which is manifested at bull-fights is absent here. Everything is accepted and witnessed with an air of boredom and martyrdom that is quite pathetic. More time is spent gazing around at the audience than at the players. Everybody carries opera-glassos, and makes good use of them.

Without doubt you would like to know how they dress; the men – who always come first, you know – wear handsome suits, displaying immaculate shirt-front and collar that would make Eastern dudes turn green with envy. Generally, the suit is entirely black, yet some wear light pantaloons. High silk opera-hats and a large display of jewelry finish the handsome Spanish man.

The ladies wear full dress, always light in color – pink, blue, pea green, white, etc. – trimmed with flowers, ribbons or handsome laces. The hair is arranged artistically, and the dresses are always cut very low, displaying neck and arms such as only Mexican women possess. Very handsome combs and pins generally grace the hair. Young girls sometimes wear flowers, but it is considered better taste to wear the artificial article, because the real are so cheap, and the former, unsurpassed by nature, command very high prices. A Mexican woman would not be dressed without the expensive fan which she flits before her face with exquisite grace. The prevailing style is a point lace fan, which adds beauty to the face and, at the same time, does not hide it from beholders, for, let it be whispered, Mexican girls are fond of being looked at. A lady considers it the highest compliment she can receive for a man to stare at her for a long time, and the men come quite up to the point of being extremely complimentary.

The prompter's box is fixed in front of the stage, and his voice is not only heard continually above that of the actors, but his candle and hands are always visible, and he often takes time to peep out and take a survey of the audience; but the Mexicans do not notice him any more than the footlights. A bell, which sounds as heavy as a church bell, rings and the curtain falls. Well, it is a sight! The managers farm out the drop-curtain to business men by the square. The enterprising advertiser has painted on a piece of cloth his place of business and curious signs. One shows a man riding a fat pig, and from out the man's mouth comes the word "Carne" (beef). How they make beef out of pork is unknown. Saloons take up the most prominent place. A house, bearing the sign "Pulque" had the side knocked out, displaying a barrel which filled the building from floor to roof. Cupid was astride a barrel, sipping pulque from an immense schooner, forgetting in his enjoyment his usual occupation of softening other people's brains with love's wine. One fat, bald-headed old fellow had gone to sleep with a generous smile on his open countenance, while from a large glass which he held in his hand the drink was running down his coat sleeve. Another fellow, equally fat and equally bald, was gazing at a full champagne glass in drunken adoration. These are a few of the curious inducements for people to patronize certain stores. The signs are only pinned on, and as the curtain comes tumbling down they fly, work and twist in the most comical style.

Naturally the spectators would grow tired gazing at such a thing, so between acts the ladies visit one another, and the men rise in their seats, put on their hats, turn their backs toward the stage, and survey the people, English fashion. They smoke their cigarettes, chat to one another, and discuss the women. The cow-bell rings again, people commence to embrace and kiss, and when the third bell rings, hats are off, cigarettes extinguished, and everyone in place in time to see the curtain, after being down for thirty minutes, rise.

Theaters close anywhere between 12.30 and three o'clock. The audience applaud very little, unless someone is murdered artistically. If a few feel like applauding other fine points, they are quickly silenced by the thousands of hisses which issue from all quarters of the house, and a Mexican hiss has no equal in the world. Ladies do not applaud, never look pleased or interested, but sit like so many statues, calmly and stupidly indifferent. After the play everyone who can afford it goes to some restaurant for refreshments. Mexicans are not easily pleased with plays; and the only time they enjoy themselves is when they have a "Zarzuela" – a cross between a comic opera and a drama. Then they forget to hiss, and enter into the spirit of the play with as much vim as an American.

Some Mexicans are quite famous as play-writers. When a new piece is ready for the boards a house is rented, and it is presented in fine style, the occasion being a sort of social gathering. Being invited, the other night, to attend one, I concluded to see what it was like. The author had one of his plays translated into English – the name now forgotten – which has met with great success in the States. I thought this would be endurable. As I entered with some ladies an usher in full dress and white kid gloves presented each of us with beautiful bouquets, and offering his arm to the ladies, escorted the party to the box with the air and manner of a prince. Once in the box, he gave us little programmes, went out, and locked the door. Interested, I watched the people as they came in and arranged themselves comfortably. Much amused and even disconcerted we were when we found hundreds of glasses turned our way and held there long and steadily, as they saw we were "greengoes," or foreigners, and with feminine timidity we thanked our lucky stars we had ventured forth without a bonnet – as no woman ever wears a hat to the theater here – so that the difference would not have been more pronounced.

At last the curtain went up, and before the actress, who was sitting on a chair, crying, could issue one blubber, dozens of bouquets were flung at her feet. Not understanding the words, the play seemed most absurd. Apparently, the girl could not marry her lover because her mother had forbidden it, as another sister loved the same man, and as he did not reciprocate she was dying; the dying sister appeared but once, then in a nightdress, and soon afterward screamed heartily behind the scenes and was pronounced dead by the actors. The men and women cried continuously all the evening, and Americans dubbed the play "The Pocket-Handkerchief." Once, when the lover told his sweetheart he was going out to fight a duel with a dude with a big eyeglass, who had loved the dead girl, she fainted on his breast and he held her there, staggering beneath her weight, while he delivered a fifteen-minute eulogy. As she was about two feet taller and twice as heavy as he, the scene was most comical, particularly when she tried to double up to reach his shoulder, and forgot she had fainted and moved her hands repeatedly. But smothering our American mirth we looked on in sympathy. How it ended I cannot tell, for at 2 o'clock I started for home and the players were then weeping with as much vigor as when the curtain first rose.

The carvings and finishing of the National Theater are superb. It is surpassed by few in the States, but the walls are smeared and dirty – no curtains deck the boxes, uncomfortable chairs are alone procurable, and, all in all, the house is about as filthy as one can find in Mexico. It is rumored that Sarah Bernhardt is to come to Mexico next December with a French troupe, and as French is as common as Spanish here, she will doubtless have large houses. It is to be hoped the managers will awaken to the fact that the house needs a scrubbing down and fumigating before that time.

As stated before, young men do not need to keep back their washerwoman's money to be able to take their best girl to the theater. A gentlemen and lady are never seen alone; even husband and wife, if they have no friends, take a servant along.

Mexico supports a circus all winter. They have an amphitheater built for the purpose, and it is the best lighted and cleanest spot in the city. It is open afternoons and evenings, except Monday. The seats are arranged theater-like – pit, boxes and balconies. Some very good performing is done, but Spanish jokes by the clowns and very daring feats on horseback are the only acts which gain applause from the Mexicans. The menagerie, for which they charge twenty-five cents extra, is not well attended, as the people can see more in the museum for nothing, and they prefer the beasts stuffed, to being stuffed themselves or stuffing another man's purse for the sight of a lion, monkey and striped donkey.

CHAPTER XIV.
THE FLOATING GARDENS.

OF course, everybody has heard of the famous floating gardens of Mexico, and naturally when one reaches this lovely clime their first desire is to go up to La Viga. I wanted to visit the gardens, and with a friend, who put up a nice lunch, started out to spend the day on the water. The sun was just peeping over the hilltops when we took a car marked "La Viga," and off we went. We spent the time translating signs and looking at the queer things to be seen. The oddest sight was the slaughter shop. The stone building looked like a fortress. Around the entrance were hundreds of worn-out mules and horses, on which men were hanging meat. They had one wagon, but the meat, after rubbing the bony sides of the beasts, was just as palatable as when hauled in it. It was built like a chicken coop, and elevated on two large wheels. On each side of the coop and lying in a large heap on the bottom, was the meat. Astride the pile sat a half-clad fellow, and in front, on the outside, sat the "bloody" driver. Trudging along in a string of about forty were men with baskets filled, with the refuse, from which the blood ran in little rivers, until they looked as if they had actually bathed in gore. We were glad when our car passed, and had no appetite for the lunch in our basket.

When the car reached its destination we alighted, and were instantly surrounded with boatman, neatly clad in suits consisting of white linen blouse and pants. Everyone clamored for us to try his boat, and the crowd was so dense that it was impossible to move. As there is no regular price, we had to make a bargain, so we selected a strong, brown fellow, who, although he pressed close up to us, had not uttered a word while the rest had been dwelling on the merits of their boats. We went with him to the edge of the canal and looked at his little flat, covered with a tin roof. White linen kept out the sun at the sides, and pink calico, edged with red and green fringe, covered the seats. The bottom was scrubbed very white and the Mexican colors floated from the pole at the end. We asked his price. "Six dollars," he answered. "No," we said; "it's too much." After more debating and deliberating he set his price at one dollar, which we accepted.

Sunday is market day, and La Viga was consequently the prettiest sight we had yet seen in Mexico. It was completely filled with boats containing produce. Some were packed full of fresh vegetables, some contained gay colored birds, which the Indians trap in the mountains and bring to market here, and others were a mass of exquisite flowers. While the man piloted his boat over the glassy waters, the ever-busy woman wove wreaths and made bouquets from the stock before her. Such roses! I can yet inhale their perfume, and how they recalled kind friends at home. Daisies, honeysuckles, bachelor buttons, in variety unknown in the States. And the poppies! Surely no other spot on earth brings forth such a variety of shade, color, and size. They are even finer than the peonies of the States.

But this boatful has passed only to bring others, ever the same, yet always new. They look at us with a pleasant smile, and we answer their cheerful salutes with a happy feeling. Along the banks we see people decorating their straw huts with a long plant, which contains yellow and red flowers. They plait it at the top in diamond shape, and not only put it on their homes, but use it to decorate the pulque shops and stretch across streets. The most disagreeable sight was the butcher at work. Every here and there along the shore are large copper kettles filled with boiling water. One man held a little brown pig down with his knee and cut its throat, while another held a small bowl in which he caught the blood. Still further up we saw the first work completed, and on sticks, put in the ground around a large charcoal fire, were the different pieces roasting. The flies were as thick as bumblebees in a field of clover, and we realized for the first time that summer, with all its pests, as well as its glories, was on our heels.

Wash day, like everything else in the labor line here, comes on Sunday. Under the drooping willows were crowds of men, women, and children. The men were nursing the babies and smoking the pipe of peace, while the women were washing their clothes. They were not dressed in the height of fashion; they were in extreme full dress – a little more so than that of the fashionable lady of the period, for none of them possess more than one shirt, and they have no bed to go to while that is being washed; so they bask in the warm rays of the sun. The nude children play in the dark waters of La Viga like so many sportive lambs on a green lawn, while the ever-faithful, industrious wife and mother washes the clothes on a porous stone and dries them on the banks – happy, cheerful, and as contented as though she were a queen.

I think I have stated before that Mexico cannot be entered except through its city gates, which are not only guarded by soldiers, but also a customs officer, who inspects all the things brought in by the poor peons and puts a high duty on them. A poor man and woman may travel for days with their coops filled with chickens, pay duty on them and have but a few cents extra for all that labor and travel. Could one blame them then if they were lazy and live on what nature grows for them without cultivation? They are not lazy, but their burden will not be lightened until this outrageous taxation, which goes to line the pockets of some individual, is removed. Even on La Viga they have the customs gate to pass. The officer examines everything, and not only charges the price, but always takes from the load whatever he wishes gratis. In one day's collection he not only has enough to run a hotel but has plenty left to sell. When a boat is packed with vegetables a long steel prong is run through them to make sure there is nothing beneath.

La Viga is from six to twelve feet deep, and about thirty feet wide. On either side it is lined with willow and silver maple trees. It starts from Lake Tezcuco, about eight miles from the city, forms a ring, and goes back to the same source. The floating gardens, so called, are found just above the Custom House. From the name we naturally expected to see some kind of a garden floating on the water; but we did not. "Boatman, where are the floating gardens?"

"There, senorita," he answered.

"What, that solid, dry land?"

"No, senorita. With your permission we will take a canoe and go in among them."

"Con mucho gusto," we replied with Harry's so-called "greaser talk," and getting into a little dugout we were pushed, at the risk of being beheaded, under a low stone bridge by our boatman, who waded in the water. We saluted the owners of a little castle built of cane and roofed with straw and went on, impatient to see the gardens.

In blocks of fifteen by thirty feet nestle the gardens surrounded by water and rising two feet above its surface. The ground is fertile and rich and will grow anything. Some have fruit trees, others vegetables, and some look like one bed of flowers suspended in the water. Around in the little canals through which we drifted, were hundreds of elegant water-lilies. Eagerly we gathered them with a desire which seemed never to be satisfied, and even when our boat was full we still clutched ones which were "the prettiest yet."

On some gardens were cattle and horses, sheep and pigs, all of them tied to trees to save them from falling into the water. The quaint little homes were some of the prettiest features; they were surrounded by trees and flowers, and many of them had exquisite little summer houses, built also of cane, which commanded a view of the gardens. The hedges or walls were made of roses, which were all in bloom, sending forth a perfume that was entrancing. The gardeners water their plots every day. On the end of a long pole they fasten a dipper, and with it they dip up water and fling it over their vegetables in quite a deft and speedy manner. No, the gardens do not float, but a visit to them fully repays one for their disappointment in finding that they are stationary.

Undoubtedly many years ago these same gardens did really float. History says they were built of weeds, cane and roots, and banked up with earth. The Aztecs had not only their gardens on them, but their little homes, and they poled them around whenever they wished. Old age, and perhaps rheumatism, has stiffened their joints and they are now and forever more stationary. Joaquin Miller said: "Now, Nellie, the gardens do not float, but please do not spoil the pretty belief by telling the truth about them." But either our respect for the truth or a desire to do just the opposite to what others wish, has made us tell just what the floating gardens really are. At the very least they repay one's trouble for the journey.

As it was about the hour for breakfast, we opened our basket and found one dozen hard-boiled eggs, two loaves of bread, plenty of cold chicken and meat, fruit and many other things equally good and bad for the inner tyrant, and last, but not least, a dozen bottles of beer. That is not horrible, because no one drinks water here, as it is very impure, and two or three glasses have often produced fever. Of course, I could have delicately avoided the beer bottles (in my articles I mean), but I could not resist relating the funny incident connected with them for the benefit of others. One of the party was a strict temperance advocate, and when the bottles were opened the beer was found to be sour, as it is a most difficult place to try to preserve bottled goods. We immediately refused to drink it, but the T. A. said he would test it, so we gave him a glass, which he drained. We were amused, but courteously restrained our smiles; but as bottle after bottle was opened, and the T. A. insisted on testing each one, our mirth got the best of us, and I burst out laughing, joined heartily by the rest. We fed our boatman, and I never enjoyed anything so much in all my life. His hearty thanks, his good appetite, his humble, thankful words between mouthfuls, did me a world of good. The sour beer which was left by the T. A. we gave him, and it is safe to say that the best of drinks never tasted as good as that to our poor boatman.

On the gardens they have put up wooden crosses and tied a cotton cloth to them; they are believed to be a preventive of storms visiting the land, as the wind, after playing with the cotton cloth, is afterward unable to blow strong enough to destroy anything. When we anchored at one of the villages, some men came down and asked us to come to their houses to eat. Each told of the good things his wife had prepared, and one, as an inducement, said, "I have a table in my house." That, of course, is a big thing here, as not one Indian in one hundred owns a table or chair. Pulque is sold very cheap at these villages, and many of the Mexicans come up in boats or on horseback to treat themselves. Along each side of La Viga are beautiful paseos, bordered by large shade trees. They form some of the many and most beautiful drives in Mexico; and on Sunday the paseos are filled with crowds of ladies and gentlemen on horseback. It is also one of the favorite places for racing, and anyone who is fond of fine riding will have a chance to see it here. Two young fellows took from off the horses the saddles and bridles, then, removing their coats and hats, they rode a mile race on the bare horses. Large bets were made on it, and everyone enjoyed the exhibition.

In the afternoon we turned our boat toward the city, followed by a boat containing a family. The father and largest son were doing the poling, and the mother was bathing her babes. She rubbed them with soap, and then, leaning over the edge of the boat, doused them up and down in the water. After she had finished and dressed them in the clothes which had in the meanwhile been drying in the boat, she washed her face and hair, combed it with a scrub-brush, and let it hang loose over her back to dry on the way to town. When we repassed the wash-house encountered going up, we were surprised to see it nearly deserted and the few remaining ones donning their clean linen, getting into their canoes and paddling around the canal. When we reached Santa Anita, a village of straw mansions, we found they were celebrating an annual feast-day, and that the town was not only crowded with guests, but La Viga was almost impassable for boats. On this special day it is the custom for everybody to wear wreaths of poppies. The flower-women, seated in the middle of the street, were selling them as fast as they could hand them out.

From a stand a brass band was sending forth its lovely strains, and beneath were the people dancing. They have no square dances or waltzes, but the dance is similar to an Irish reel – without touching one another, and merely balancing back, forth and sideways. Pulque was flowing as freely as Niagara Falls, and for the first time we realized what "dead drunk" meant. One woman was overcome, and had been drawn out of La Viga into which she had fallen. She lay on the bank, wet, muddy, covered with flies, face down on the earth, with no more life than a corpse. She was really paralyzed.

After we tired of watching them we continued our journey, our boatman wending his way deftly between the crowds of others who were making their way to the feast. They all greeted us and said many pretty things, because I had put on a wreath. They considered I had honored them. Nearly every boat had one or more guitars, and the singing and music added a finishing touch to the already beautiful and interesting scene. About 200 mounted and unmounted soldiers had gone out to keep the peace, but they entered into the spirit of the thing as much as the others, and doubtless would consume just as much pulque before midnight. Hailing a passing carriage, as we landed, we drove to our house, jotting down the day spent on La Viga as one of the most pleasant of our delightful sojourn in this heavenly land.

CHAPTER XV
THE CASTLE OF CHAPULTEPEC.

WHEN Maximilian first established his royal presence in Mexico he began to do what he could toward beautifying this picturesque valley. The city had been rebuilt on the old Aztec site – the lowest and worst spot in the land. Maximilian concluded to draw the city toward a better locality. In order to do this, he selected Chapultepec as the place for his castle, and built lovely drives running from all directions to the site of his residence. The drives are wide, bordered with tall trees, and form one of the prettiest features in Mexico. The most direct drive from the city is the paseo, spoken of in a former letter as the drive for the fashionable. Maximilian intended his home should be the center of the New Mexico, and the paseo – "Boulevard of the Emperor" – was to lead to the gate of his park. From the Alameda to Chapultepec the distance is 5450 yards, with a width of 170 feet. The paseo contains six circular plots, which Maximilian intended should contain statues. Strange to say this plan is partly being executed. Some already contain an equestrian statue of Charles IV., claimed to be second only to one other in the world; a magnificent bronze statue of Columbus, and they are erecting one to Guatemoc and one to Cortes. On either side of the paseo are grand old aqueducts, leaky and. moss-covered, the one ending at the castle, the other going further up into the mountains. One is said to be nine miles in length. These aqueducts hold very beautiful carved pieces and niches, every here and there, in which are placed images of the Virgin.

Terminating the avenue rises the castle, on a rocky hill some hundred-feet high. The castle covers the entire top and stands like a guard to the entire valley. Many hundred years ago the King of the Aztec Indians had this for his favorite palace. Here he ruled, beloved by all, until the white-faced stranger invaded his land, outraged his hospitality and trust; stole his gold and jewels and replaced them with glass beads; tore down his gods and replaced them with a new; butchered his people, and not only made him an imbecile, but caused him to die at the hands of his once loving subjects the despised of all the people. Poor Montezuma! the wisest, best and most honorable King of his time, after all his goodness, his striving for the light of learning, to die such a death.

Since Montezuma wandered beneath the shades of Chapultepec – "Hill of the Grasshopper" – it has been the chosen resort of the successive rulers of Mexico – the theme of poets, the dream of artists and the admiration of all beholders. A massive iron gate, guarded over by dozens of sentinels, admits you to a forest of cypress which excels anything on this continent. The grand old trees, many centuries old, are made the more beautiful by the heavy dress of gray moss which drapes the limbs. The broad carriage road, to which the sun never penetrates, and where the beautiful, shadowy twilight ever rests, winds around and around until it gains the summit. The old bath of Montezuma stands a lovely ruin in this lovely grove; above it is built an engine house for the waterworks, which are to supply the city instead of the aqueduct. With regret we gazed on it, the only blot on the otherwise perfect paradise, and wished that someone, with the taste of Maximilian, had interfered before this mark of progress had been decided upon.

The silvery lake, alive with geese and ducks, and bordered with lilies of the Nile and other beautiful flowers, nestles like a birdling in the heart of the greensward. The fountains play and sing their everlasting song, while birds of exquisite colors mingle their sweet melodies with the tinkle of the falling waters. Plots of flowers vie with each other to put forth the most beautiful colors; all nature seems to be doing its utmost to show its gratitude for being assigned to this beautiful spot. Far back in the forest, is a smooth, level place, where moonlight picnics are often held. The soft drapery of Spanish moss hangs low, yet high enough not to interfere with the headgear. Beneath its shadows one would fain forget the world. We no longer wonder at the "manana " of the natives, and can clearly see why they wish to live as slow and as long as possible.

When Montezuma reigned supreme he was accustomed to gather together his wise men, and while sitting beneath the shade of a monstrous cypress they would discuss the topics of the day. For this reason, the tree is named "The Tree of Montezuma." It is said to be two hundred feet high and sixty feet in circumference. It is heavily draped with moss, and is the most magnificent monument any king could have.

Half way up the hill is an entrance, almost hidden by moss and other creeping foliage, which leads into a cave. The first chamber is a very large room hewn out of the solid rock. At the opposite side is an iron door, barring the way to the cave proper. Many different stories are told of it. One is that the cave was here before the time of Montezuma, and that untold wealth has been hidden in its unexplored recesses when different tribes went to war. Another says that when Cortes was forced to leave he buried his ill-gotten wealth in its darkened depths. The less romantic story is that the subterranean sally-port, which leads down from the garden on the roof of the castle, opens into the cave; they once tried to explore it, and found within a mammoth hole. A rock thrown in was not heard to strike the bottom, and even the bravest feared to go further. The rocks on the hill are covered with hieroglyphics, which archaeologists have not succeeded in translating; the brick fence around the winding drive has passed its day of beauty, and the posts alone remain of the lamps which once lighted Maximilian's pathway.

Having obtained a ticket of admission to the castle from the governor of the National Palace, we took a party of tourists with us and proceeded to investigate. When we had mounted the hill, and walked through the iron gate into the yard, the uniformed sentinel called out something in Spanish, loud and long, and a drummer boy quite near beat a hasty roll. "They must think we intend to storm the castle," said one of the ladies in evident alarm, but her fears were quieted when a young cadet came from the building and offered to show us around. "Can you speak English?" I inquired. "No. I will find someone," he answered in Spanish, and off he went. However, we lost no time waiting for his return, but went to the door of the castle and handed our pass to the guard. "Momento," he said, and he also disappeared, but only to come back accompanied by a handsome, middle-aged officer, who told us, in broken English, our pass was good, and while the guard would take us through the castle he would get us another escort for the rest.

The castle is being renovated for a Mexican White House. A New York firm is to finish it at a cost of one hundred and sixty-five thousand dollars. Our disappointment increased as we roamed through room after room to find all mementos of Maximilian and Carlotta destroyed. Even what had been their bedchamber was a total ruin. The only things that remain are three poor pictures on the wall facing the garden. They had been spoiled, and before many hours the last thing to recall the murdered emperor and the blighted empress would be totally effaced. President Diaz is to move here when the repairs are finished; but if they are no faster with the work in the future than they have been in the past, what they have begun will be old-fashioned before the rest is completed, and Mexico will have added two or three more names to its list of presidents.

On top of the castle is a beautiful garden, full of rare plants and handsome trees and shrubbery. Fountains are plenty, and statues of bronze and marble are strewn around in profusion. The stairway is made of imported Italian marble, and the balconies of alternate blocks of Italian and Puebla marble. The effect is superb. The famous sally-port leads down through the castle from the center of the garden. It is fenced in around the month with a brass railing and covered with green vines. Magnificent aquariums divide the flowers at intervals, and the little gold and silver fish play about in the water as if life was all joy. When one looks around the beautiful landscape, the romance of the historic past fades before the grand reality of the present. From this majestic spot one commands a view of the entire valley – the soft, green meadows, the avenues of proud trees which outline the gray roads that always fade away at the foot of the chain of mountains which encircle the valley like a monstrous wall. The faint blue and purple lines of the mountains appear small and insignificant when the gaze wanders to those two incomparable beauties, Popocatapetl and Ixtaccihnatl. All nature seems a prayer. Grand old Popocatapetl stands with its white, snowy head at the feet of the White Lady. Perhaps nature has assumed this tranquilness while awaiting the old, white-headed man to say the last sad words over that beautiful still form.

At the back of the castle is the Military Academy, or West Point of Mexico. Three hundred cadets, with their officers, are housed here. The school is kept in the best of order, and when the cadets finish their seven years' course they are well prepared for future duties. The cadets belong to the best families and number a lot of handsome men. The stairway which divides, or rather connects, the two buildings is an odd yet pretty structure. It is built in an arch to the height of ten feet. Then starting out in opposite directions are two other arches, which connect the buildings. These arches – the stairway, of course – have no supports whatever, and one is almost afraid they may cave in with their weight. When they were finished someone remarked to the builder, "They will fall down if one man mounts them." "Bring a regiment and put on them, and I guarantee they stand," replied the builder. This was done, and they were found to be as firm as a mountain. They are certainly one of the prettiest pieces of architectural work ever executed.

In the library of the academy are oil paintings of the cadets who fell in defense of Chapultepec. They were handsome young boys, and a fine marble shaft, enclosed with an iron fence at the foot of the hill, is erected in commemoration of their heroic deed. The prettiest boy of the lot, with sunny locks and blue eyes, folded the flag, for which he was fighting, to his breast, and stood with a smile on his face while his enemies cut him into pieces. He was but thirteen years old. His picture occupies a prominent place, and beneath it stands the flag, dyed a dark crimson with his heart's blood. The cadets keep those little heroes' memories green. Every morning they place wreaths of flowers on the monument as they march on their way to the meadows below to drill.

The cadets have two queer pets, a wild pig and a monkey. The latter is their companion. He performs in the gymnasium with them, and does some wonderful feats. He is truly a smart, cunning little fellow, and exhibits much intelligence. He is fond of the boys, and the boys return his affection. When they come to town on Sundays they never forget to take some sweetmeats back for him; and he never forgets to expect the treat, and he gets very loving and confidential about that time. He hugs the returned youth, and prys into his pockets with as much enthusiasm as though he had been absent for months. Every cadet has a bed with his name, number, etc., on it. A combination desk and wardrobe stands by the side, and in the bottom is a tin pan. At 5.30 they arise, and when the order is given they take up their tin pans and march out to the side of the building. From a large basin they take the water, and placing their pans on a stone bench many yards long they wash themselves. On Sundays they can go to bull-fights, to town to see their relatives, or do anything they wish, unless they have neglected their studies the week before, when they are kept at school for punishment. They are taught French, Spanish, Greek, and English. They are extremely polite, and have not the least objection to flirting. Though they are short in stature they have good forms and are splendid horsemen. In fact, they are the beau ideal of any girl who likes embroidered uniforms and brass buttons, topped off with that cavalier style no female can resist.

CHAPTER XVI.
THE FEASTS OF THE GAMBLERS.

THE Mexicans, as a people, have an inordinate passion for gambling. They gamble on everything. Poor peons have been known, when their money was gone, to take the rags off their backs and pawn them in order to get a few cents to lose. Men possessing thousands have gone into houses at night to be hauled away in the morning a corpse, without a dollar to pay funeral expenses. Gambling reached such a stage that the government saw it must interfere. Consequently, they prohibited all street gambling and started lotteries, in which prizes are drawn every other day. The main prizes range from $500 to $5,000. Crippled, blind, aged, poverty-stricken men and women are on the streets at all hours selling numbered strips of tissue paper marked "Lottery." The seller wears a brass badge in the shape of a half-moon as proof that he or she is employed by the government. No trouble is experienced in selling the tickets, as everybody buys, foreigners as well as Mexicans. The tickets range from twelve cents to twenty-five dollars. When the drawing is held a printed list showing the fortunate numbers is posted in the court. People of all nationalities and in all stages of dress crowd around the notice. Many turn away unhappy, while some few smile over their gains. It is said the proceeds are devoted to useful and charitable works. The income, at any rate, must be a princely one.

Gambling houses are also run on a large scale. They are licensed by the government. Once every year, in the month of February, gamblers procure a license and open houses at Tacubaya. During these four weeks all are allowed to gamble here in any style they wish. This chance picnic is called "the feast of the gamblers." At three o'clock every afternoon ladies in carriages, men on horseback, the poor in the street cars, all bound for the one destination – Tacubaya – present a beautiful sight. From the energy displayed, the hurry to pass one another, the evident desire to get there first, one would think it the first holiday they had had for years, and all were determined to get the most out of it!

To reach the scene the tourist must take a two-mile drive along a wide road, bordered on either side with trees of luxurious growth and shade, beneath which beautiful, pure-white calla lilies and scarlet-red geraniums lift their pretty heads in the perfect abandon of naturalness and liberty. Dotted here and there over the lovely valley are green fields, adobe huts, and whitewashed churches, with superb Chapultepec ever in view, as a crown or guard to the vast valley beneath. The gates of Chapultepec, with its sentinels and mounted guards, are passed, and in a few minutes more we are in Tucubaya.

"We will have to alight here," said our guide. "The streets are so full it is impossible to drive through."

Impossible to drive; it was almost impossible to walk. As we stepped from the carriage several peons, who had come to meet us, knelt on the ground and spread out their serapes before them, displaying a few silver dollars, big copper one and two cent pieces and three cards; the cards were deftly crossed, face downward, one after another, with astonishing rapidity, while the "tosser" kept singing out some unintelligible stuff, apparently, "Which will you bet on?" Quickly a peon steps forward and lays a $10 bill on one card. The "tosser" shuffles again, the man wins and puts many silver dollars in his pocket. This excites the watching crowd, which presses forward, and many women and men lay down their money on certain cards, only to see it go into the pile of the "tosser." One failure does not discourage them, but they try as long as their money lasts, for it is impossible to win. The "tosser" has one or two accomplices who win the first money to excite the crowd or again to increase their waning energy.

The "tosser" and his accomplices will follow Americans, or "greenoes," as they call us, for squares. When you pause they prostrate themselves before you; the stool-pigeon always wins and tries to induce the stranger to play – even pinches off the corner of the card, saying "It will win; bet on it;" "Senor, try your hand." "Senorita, you will be lucky," whispers the accomplice as he gazes at you in the most solemn manner. Wild-eyed women, who smell strongly of pulque, with disheveled hair and dirty clothes, beg for money to try their luck.

Each side of the street is filled with tents. In the center and along the houses are women squatted on the ground nursing their babies and selling their wares, which consist of everything ugly. Some build little charcoal fires, above it, suspend a flat pan, and on it fry some sort of horrible cakes and red pepper, which are sold to the gamblers. At the foot of a large tree sat an ugly, dirty woman. From a big earthen jar by her side she dealt out pulque to the thirsty people; the jar was replenished repeatedly from filled pig skins. At another place tomatoes and salad were laid out in little piles on the ground. A little naked babe lay asleep on a piece of matting, and a woman was busy at the head of another – not reading her bumps, but taking the living off the living – and she did not have to hunt hard either. Similar scenes repeated themselves until one longed for something new.

The restaurants were numerous. A piece of matting spread on the ground constituted the tables, with the exception of three old wrecks that could hardly stand. Cups of all shapes, but none whole, lay claim to being the only dishes in sight. Large clay jars, tin boilers, etc., were the coffee urns.

Among all the mob that gathers here, a fight is an unheard-of thing. "It is old California repeated," said Joaquin Miller, "with the rough people left out." Rough, in a certain sense, they are, and ignorant, yet far surpassing the same class of people in the States; they possess a never-failing kindness and gentleness for one another; the police carried one woman who was paralyzed from pulque as tenderly as if she were their mother, while a sympathizing crowd followed; two peons supported between them a pulque victim, who was so happy that his spirits found vent in trying to sing a hiccough song. Another peon, only half sober, got his drunken companion on his back and trudged off, in a wavering manner, for his home.

In the tents along the street a second class of people gamble. Some tables have painted on them three faces – a red one, with a white and green one on either side – on which the men gamble. Musicians with string instruments furnish pleasing airs, and women in picturesque costumes do the singing and dancing. The most popular song is "I am a pure Mexican, no Spanish blood in me." The people scorn the idea of Spanish blood, and boast of being of pure Indian descent.

Over the top of high walls peep the green trees, and the vines crawl over, hanging low down on the outside. Enter the vine-draped gateway and you will see a garden as fine as any city park. A smooth walk leads to all sorts of cunning little nooks; large trees spread out their heavy arms; the perfume of thousands of beautiful flowers scents the air; playing fountains mingle their music with the exquisite melody of the string bands placed at intervals throughout the grounds; statues glisten against the green foliage; well-dressed men and finely clad women are visible on every spot – everything animate and inanimate adds to the picturesqueness of the beautiful scene.

In the buildings, which are decorated outside with pictures from happy scenes in life, are tables and chairs, the walls being hung with fine paintings and expensive mirrors. On the green table-cloth is placed $10,000 and $20,000 – the former sum on the roulette table, the latter on the card board.

The money is half gold and half silver. Before the hour of playing these tables are left unguarded; people go in and out at pleasure, but all are too honorable to take one piece. Ladies and gentlemen sit or stand around, smoking their cigarettes and betting. One woman lost $500 in a few moments, but her face never changed. A man stood at a roulette table, and, commencing with $10, was in a short time the possessor of $750. He never changed countenance, and after getting the "pot" together he exchanged it for greenbacks and walked off. Any one playing can order what they wish to drink at the expense of the proprietor. Fine restaurants are also run in connection with the establishment.

One gambling hall is hung with Spanish moss in the shape of a tent, which reflects in the mirrors forming the walls. It is beautiful and reminds one forcibly of what fairyland is supposed to be. Every large house has a notice posted informing patrons that they furnish, free of charge, conveyances for the city at late hours. One man almost broke the bank and had to get a wagon to haul his money to Mexico. Others won $5000, $10,000 and $20,000, but notwithstanding this one house made $200,000 the first ten days. Electric lights enable the players to keep the game up all night, and unique torches furnish just enough light in the gardens to show the way and fascinate the sentimental.

Tired at last, we wandered forth and visited the beautiful old cathedral which all Mexican towns possess, walked through several plazas and examined the fine fountains, flowers and monuments, and at last traveled to the top of the hill in order to view the country around about. Seated on the eight-foot bank of the military road, we watched the Indians going to and from the city. First came a drove of burros walking quite briskly, as if they feared the load left behind might catch up and insist again on being carried. A number of women wrapped up in a straight piece of flannel and a piece on their heads in the style of the peasant girl in the "Mascot," passed by. On their back were huge bundles of wood and scrubbing-brushes. "Buenas noches, senora; buenas noches, senorita; buenas noches, senors," they cried out pleasantly as their bare feet raised enough dust to encircle them. Their black eyes gazed on us in a friendly manner and their lovely white teeth glistened in a cordial smile. "Poor human beasts of burden! Give the little one some money," we whispered. "Here, this is yours," he called, in Spanish, holding forth a silver dollar. The smile faded from her face. "Gracias, no, senor!" and she quickly passed on, too proud to accept what in all probability was more than she ever owned.

The sun had long gone down; dark clouds draped the "White Lady;" Chapultepec looked dim and hazy. With regret we left our prominent position, passed the handsome palaces of Escandon, Mier y Celis and Barron, walked through one of the handsomest villages in Mexico – Tacubaya – and in a few moments reached our carriage, homeward bound, leaving the "Feast of the Gamblers," just in the height of its glory.

CHAPTER XVII.
FEAST OF FLOWERS AND LENTEN CELEBRATIONS.

IF they had put both in a kettle and, after constant stirring, poured the contents out, there would not have been more of a mixture of religion and amusement than there was during Lent; to a sight-seer it looked as if the two forces were waging a battle to see which would predominate. It was very interesting, more so from the fact that in no other place on earth is Lent celebrated like it is in the City of Mexico. I think I told you how the carnival season opened, with balls, picnics, and driving in full dress on the paseo; then suddenly everything collapsed, and the city put on somber robes. Bells tolled forth from morning until night, and every other day was a saint's day, when, Catholic or otherwise, we were compelled to fast; the stores closed, and everything came to a standstill. All the night previous fireworks were set off, and revolvers cracked until one's wildest wish was that their inventors had never been born.

One morning I was surprised to learn I could not have any coffee – the solitary cup which constitutes our dainty, delicious breakfast here. My limited Spanish prevented my giving vent to my feelings, and so I nursed my righteous wrath while I took observations. The whole house was closed and darkened, the mirrors were covered with purple cloths, and every little ornament, which had hitherto decorated the house, was missing. All the people of the household were dressed in black, talked in whispers, and walked around on their tiptoes. Dinnertime came and we sat down to a bit of dry toast (butter is an unheard-of thing), black coffee, chile, or red pepper, and beans. By this time, I began to get "shaky," especially as they did not talk and pulque was dispensed with. After saying: "Someone must be dead;" "They must have gotten into some kind of trouble, and are trying to make believe they are away," I decided to quit "guessing," and try to find out the true cause of these strange doings. Finally, I decided to see if any of my Mexican "bears" were visible; and, going through the parlor, I opened the window leading to the balcony. Just as I had removed all the monstrous bars, my landlady came rushing to me, with a burning candle in one hand and beads in another, and in louder tones than she had spoken before she besought me not to open the window. Completely mystified and feeling sure they had done some terrible deed, I closed the bars, with one longing sigh to my "bears," and. then catching her by the shoulder, asked, in trembling tones: "Tell me, what have you done?"

"No comprehende," she ejaculated, looking at me as if I had lost my senses.

"Porque?" I asked, pulling her around, and pointing to the bare tables and cabinets, the draped mirrors, the barred shutters.

"I am sad because it is my saint's day and my mother's day," she explained, and she took me into her room, where everything was draped in somber colors. Below the picture of her mother were a number of burning candles placed around a large cross. Before this cross the rest of the family were on their knees, and as I slipped out and closed the door I saw her sink down beside them, with a look of submission on her face. I have nothing more to say, except that I am glad that before a similar day rolls around I shall be over the Rio Grande and doubtless at home.

Holy week began on Piernes de Dolores (Friday of Sorrow), April 16. As early as 3.30 in the morning the bells began to toll, and people flocked to the churches. At five o'clock we started for La Viga, where this day is celebrated by the Feast of the Flowers, or Paseo de las Flores (Flower Promenade). Even at that hour the way was crowded with people laden with flowers. When we reached La Viga we found it filled with canoes and boats burdened with beautiful flowers of every description. Aa far as we could see up La Viga it was the same – picturesque people paddling their equally picturesque boats in and out and around the crowd. Some of the boats were ready for hire. They had awnings made of cane covered with ferns and flowers. Very few could resist their inviting appearance, and by nine o'clock there was not an empty boat to be found.

Along the fragrant, grassy banks sat flower girls surrounded by heaps of ferns, creamy lilies, delicious pinks of hundreds of shades, geraniums and fuchsias of wonderful size and color, and roses whose colors, sizes and perfumes bewildered me. Honeysuckles, roses, lilies and poppies were woven into wreaths, which people bought and wore on their heads and around their shoulders. Eating-stands were about as plentiful as the flowers, and everything that was ever made in Mexico was here for sale. They did a big business, too. Gay crowds would sit down on the grass and take breakfast off of a straw petate as merrily as if in the finest dining-room. Some of these booths were fixed up with canvas covers and flower sides; other long booths were fitted up in the same manner, hung with the Mexican colors and filled with chairs, where the tired could pay a medio (six and one quarter cents) and sit down. Three bands in holiday attire sent forth lovely strains, alternately, from similar booths; the trees on either side kept the paseo shady. It was filled with people riding and driving; the riders, who numbered many ladies, formed a line in the center and the carriages drove around and around, down one way and up the other. Most of those out driving alighted and mingled with the masses. It was certainly a most enjoyable scene.

At several places we found things for sale which looked like dahlias, with a strange mixture of colors. None could determine just what they were, but presently we found a man and woman manufacturing them. They were nothing more or less than long radishes, which with his penknife the man turned into all kinds of flowers, as well as crosses and other designs. The woman delicately touched one part one color, another another, until they formed one of the most beautiful of the many strange sights on La Viga. There was quite a rush for them, and the happy purchasers triumphantly carried them off, while the less fortunate looked on with regret. I got a number, but before the next morning their beauty had departed forever, and their perfume was loud and unmistakable. Of course, there were plenty of venders and beggars there. The venders had wax figures representing ballet dancers, rope-walkers, angels – any sort of female that was skimp in her wearing apparel. Others had men fighting bulls, monkeys on horseback, baby dolls made of rags, and every little thing which could be invented.

This feast lasted until Sunday evening, and there was not a moment from three o'clock Friday morning, until twelve o'clock Sunday night, but what the place was crowded worse than Barnum's show in its brightest days. The prettiest sight was when the people returned to town. Every carriage, even to the driver's seat, was filled with flowers. The horses and riders were decorated with wreaths, and in this manner, they all returned to their homes. I must describe one rider to you before I leave La Viga. He rode a beautiful black horse. The Mexican saddle was a bright, deep yellow, covered with silver ornaments, and a bright sword dangled at the side. The bridle was entirely of silver, even to the reins, and silver cord and tassels decorated the horse's neck. The rider's pants were black and fitted as if he had been poured into them. A row of silver buttons, at least the size of pie-plates, reached from waist to knee, where they were met by high side-buttoned boots. An immense silver spur completed that part. His vest was yellow velvet, his coat blue, and his wide sombrero red, all heavily trimmed with silver, while at the back, peeping beneath his coat, were two mammoth revolvers. He was the most gorgeous butterfly I ever saw, and attracted attention from Mexicans as well as myself.

Sunday was observed by the churches as well as on La Viga. It was Palm Sunday, and the Indians had made pretty things out of dry palms which they sold to the people for from a real (twelve and one-half cents) up to cinco pesos ($5). The devout took these to church and had them blessed, and after carrying them home they were fixed to the bars of windows, the balconies and above the doors, where they will stay for the whole year. They say they keep the devil out, and that is their reason for using them.

Excursion trains were run in from all the connecting points, people appeared in the most gorgeous hues, and venders had no trouble in selling the effigies they carried. Holy Thursday came and the bells tolled from early morning until ten o'clock, when everyone was silent in sorrow for the crucifixion. Mass was said in the morning, and all turned out to attend divine service. In the Alameda, Zocola and paseo bands, to the number of three or four, delighted their hearers. It seemed rather strange to stand within the church door and hear the voice of the priest repeating mass, the piano playing a soft prelude (no pipe organs are permitted during holy week), and the band mingling the lively strains of some light opera, or something equally ridiculous, with this solemn service, the altars were all hung with squares of silver or gold tinsel, which were constantly in motion. Thousands of candles lighted up the gloomy building, and Christ and the Virgin were the only images in sight. At 3 o'clock in the afternoon they brought in what they said were the oldest and most neglected of beggars. The priest washed their feet, and after making the sign of the cross with holy oil upon them, they were allowed to depart. I noticed these men's feet had been washed recently, and also that there were dirtier and poorer people in the audience. However, the washer took good care not to touch the feet without an intervening towel. At night the churches were brilliantly illuminated. It would be hard to give an estimate of the candles required, but I fully believe that in some of the big edifices 20,000 would not be a bad guess. The devout were all on their knees, and everything was as silent as death, except the piano, which still kept up its soft, soothing melody.

On Good Friday all the men and women were dressed in black, and every church was draped with purple. The Virgin was dressed in heavy black velvet. The poor Indians laid flowers, money and candles around the image, and they could not have been more deeply touched had the crucifixion taken place then instead of so many hundred years ago. They kissed her feet, her garments, and the floor before her, and showed in a thousand humble ways their love and devotion.

The ceremony of the Tres Horas (three hours) was celebrated in Texcoco. First a lot of masked men ran around the yard with sticks, beating the bushes, trees and flowers as though in search of something. Then one of the men who was far from representing Christ in form, feature or complexion, took a heavy wooden cross on his shoulders and walked into the church, being lashed with a leather strap by the masked men. When he fell the people covered their faces and groaned. He fell three times before reaching the altar, where an effigy was nailed to the cross. The sounds of the hammer and groans and cries of the people made one feel as if somebody had dropped a piece of ice down their back. Finally, amid the most heartrending cries, the cross was raised and the ceremony was over.

All day wagons, horses, boxes, everything in the toy line, with a racket in them, were sold to the people. All the venders were located around the cathedral and Zocolo, and the din could be heard several squares away. These are called matracas. When Christ was on earth, they say, they had no bells with which to call the people to mass, so these matracas were made, and a number of men would promenade the streets, swinging them around to keep up the incessant cracking. The men would cry out, "The hour has come for mass, the hour has come for mass," and the faithful would hurry away to count their beads and say their prayers. A foreigner told me this custom was still in vogue in some parts of his country, France, during holy week. Hideous effigies, called Judas, were for sale. Little ones made of lead were bought and tied to the button-hole, the parasol, the bracelet, the belt, or any other convenient place. Some made of plaster of Paris and paper, from three inches to twelve feet long, were bought by old and young and carried home for Saturday.

Sabado de Gloria (Saturday of Glory) came bright and sunny. All along the streets were strung long Judases, some having pasted on them the thirty pieces of silver for which he betrayed Christ; the image was made in the most horrible form – as a negro, devil, monkey, half beast, half human, every form that could possibly be thought of. At 11 o'clock the bells began to ring merrily, as though rejoicing over the fate of Judas, and a match was applied to every image in the town; they were all filled with powder, and with one accord there was a universal bursting and tearing and rejoicing throughout the city. As fervent as had been their devotion to the Virgin, just as strong was their hatred of Judas – even the smallest scraps they tramped upon.

By 12 o'clock gay colors were resumed, carriages which had been rigorously kept out of sight came forth and were flying down the paseo as if glad that the time of quietness was past. All places of amusement, which had been closed during Lent, began sticking up posters announcing a grand opening on the next (Sunday) evening. The noise of the matracas grew fainter and fainter, and gradually ceased. The wind picked up the stray pieces of Judases, played with them awhile, and then carried them out of sight. The venders who had jammed the Zocalo gradually disappeared, the music in the different parks ceased, and Lent seemed as far gone, by the time 12 o'clock rang forth, as though six months had passed. Such is life.

On Sunday the theaters, bull-fights, circus and race-courses were well attended. The bull-fights were advertised as the last of the season. The one I attended was excellent. The bulls were good ones, and some very new and striking features were introduced. One man sat down on a chair in the center of the ring with two banderillias in his hand. The door was opened, and the bull rushed in and at him. He sat there, and as it put down its head to gore him he stuck the banderillias into its neck and sprung aside, while the bull knocked the chair into atoms. Everybody cheered, and threw the fellow money and cigars. After this toro had been dispatched, one man lay down on the ground and another stood over him, keeping his head between his legs. Again, they opened the door and let a toro in. It rushed for the men, but the one standing stuck the banderillias into it with such force that it roared with pain and took after one of the other fighters in the ring, leaving the two men unhurt. The very daring of this delighted the people, for if the man had missed the bull both of them would have been killed without the least trouble.

One toro had horns about four feet wide, and at the first plunge it killed one horse. Then it caught another horse and threw it on its back, the rider underneath. The fighters tried to draw it off, but it stayed there until the horse was dead. All that could be seen of the rider was his head, which he tried vainly to shield with his arms. They carried him off for dead. This toro was very hard to kill. It required seven lunges of the sword to convert him into beef. One toro refused to fight, and when stuck with a sharp pica he jumped over the fence and was with the audience. Such a scrambling! Most of the people threw themselves into the ring, about the first ones to go being the guards, who are placed around to take care of the people. It was quite a while before quiet was restored, and the toro lassoed and removed.

Bull-fights have lasted longer this season than ever before, as it is impossible to fight during the rainy season. Now a man comes forward and says he is going to cover his ring and have fights all summer; this will make the light in the ring dim, and the fighters will be at a disadvantage, not being able to calculate their distances. It will also make the fights more dangerous and more interesting. It is needless to add that the people are delighted at the prospect. Last Sunday one man got so excited over the big toro's fighting that when it was to be stabbed he got down into the ring and, taking off his high silk hat, asked the judge's permission to do the work. The audience rose to their feet and shouted "Yes, yes," but the judge was unkind enough to refuse, and thereby deprived us of seeing a fellow in broadcloth gored because he thought he could kill a toro.

Congress is in full session now. The other day they passed a bill which was strongly opposed. It is to the effect that anyone caught meddling with the railroads will be shot down instantly without a moment's warning, and without a trial. Doubtless many will say that it is a first-class law when they think of the wrongs committed on the railways in Mexico. But it is such a law as will allow thousands of Mexicans whose "honor desires satisfaction" to take advantage of it. The victim is shot, and after he is dead the shooter steps forward and swears that he saw him meddling with the railways, or knew he had designs on them. This is all he has to do to be freed of the murder. While we believe in dealing out unmerciful punishment to train wreckers, yet this law is fit only for uncivilized countries, and least of all for Mexico, where people shoot on the least provocation, ofttimes just for amusement, or to test their unerring aim, piercing the brain or heart every time. It is, certainly, a grand chance for those who have a desire for revenge to obtain it and go scot-free.

However, the law is only to be tried for one year, and if it proves good it will be adopted permanently. Now is the time for those who claim the country is ruined by a ring to remove some of its links, especially the key and padlock, and by doing so once again proclaim liberty, and prove to the people that the "shoot without trial law" really did some good.

Cinco de Mayo (5th of May) was the next big day for Mexico. Then they commemorated the victory over the French, and it is done in princely style. A French paper rather sensibly remarked that it would look better if the Mexicans, dropped this foolishness, as the French whipped them on the 4th and again on the 6th. Some little government-paid sheets came out in editorials as mad as turkey gobblers at the sensible insinuation.

I for one am glad Lent and its eggs, red-pepper, and bad-smelling fish is gone. What cowards our stomachs make of us all. I really have begun to long for home, or rather home-cooking. I have made out a list which I view every day, and see how much longer my stomach will have to endure this trash. Fixty-six more mornings to drink black coffee and long for even ham and eggs, with heavenly thoughts of hot cakes and butter. Fifty-six more noons to eat boiled cheese, meat stuffed with chili (red pepper), fish boiled in chili, with the fins, head, eyes, and tail still adhering, dolce (dessert) of fried pumpkin sprinkled with chili; fifty-six more suppers to eat the same bill of fare set up cold; fifty-six more evenings to wonder why pulgras and chinches were ever invented. By the way, if it were not for their musical names they would surely be unendurable. There is a great deal in a name, after all, and if I had to call them fleas and bedbugs I should take the next train for the States. Well, I have fifty-six more nights to spend in an iron-bottomed bed and then I shall cross the Rio Grande, and try once again the pests which inflict mortals there.

CHAPTER XVIII.
GUADALUPE AND ITS ROMANTIC LEGEND.

WE went up to the Zocalo to take a car for Guadalupe. All the street cars start from this center, and on some lines trains of three to ten in number are made up, so that they may be able to resist the bandits who sometimes attack them – at least, so the corporation claims. We determined to try a second-class car, in order to find out what they were like. Our party seated ourselves and watched the crowd as they came surging in. Two big fellows, dressed in buckskin suits and wearing broad sombreros, who sat opposite, never removed their gaze from us. A pretty little girl and an old man who sported a hat about two inches high in the brim, deposited themselves on one side of us, and a black, dried-up old fellow occupied the other.

When the car was about filled, a woman with a baby in her arms, followed by her mother and husband, came in; the women sat down facing us, while the husband, who wore a linen suit – pretty dirty, too – and carried a large purple woolen serape, of which he seemed very proud, wedged himself in between us and the piece of parchment on our left side. We were inclined to resent this close contact, and were beginning to regret we had not taken the other car, where the people are a shade cleaner, when a lot of Indian women, with babies and bundles, crowded in, and, with a sudden rush which knocked the standing ones on to the laps of the others, we were off at a 2:40 gait. The women sat down on the floor of the car, except one who was dressed a little better than the others. She came up to the dirty Indian by my side and told him to get up. He was about to do so as an utterance of thanks escaped our lips, when his mother-in-law and wife commanded him to sit down again.

This he did in all humbleness, but the woman in black commanded him to rise, as he had no money to pay his fare. His mother-in-law's ire was up, however, and she ordered him to display his wealth. He took out a handkerchief, untied the corner and displayed one silver dollar and some small change; then the old lady dived into the bosom of her dress, and untying a similar handkerchief, displayed her worldly all. The woman in black was convinced she had struck the wrong man, so she sat down on the floor and related her side of the story to the people in her end of the car, while the mother-in-law dealt out the same dose at the other end. The conductor came in, and, straddling over the women on the floor, sold the tickets for six and a half cents. Another conductor followed to collect the same, and soon we reached our destination.

Guadalupe is the holiest shrine in Mexico. It is the scene of a tradition that is never doubted for an instant by the people. In 1531 the Virgin appeared one evening to a poor peon, Juan Diego, and told him to go to some wealthy man and say it was her will that a church be built on that spot. The Indian, in a great fright, obeyed her command, but the wealthy fellow refused to put credence in the incredulous story, so the peon returned and told the Virgin, who was still there, of his failure. She told him to return and show his tilma (apron) as proof.

The amazed fellow did so, and the light disclosed the picture of the Virgin painted on the apron. Still the unbeliever doubted, and the Virgin sent for the third time a bunch of fresh roses such as never before grew in this country. The infidel took the flowers, and the picture of the Virgin fell from the heart of a rose. He was convinced, and built a large church on the spot where the Virgin appeared.

The church is a fine one, decorated with statues, paintings and gold. The silver railing weighs twenty-six tons, and is composed of a metal composite. The church authorities have received numerous offers for this rich relic. Some persons desired to replace the railing with one of solid silver, but this bargain was not accepted. Diego's apron is above the altar in a frame. On it is painted a picture of the Virgin, but, to say the very least, it was not drawn by a master hand. The bunch of roses, which, they claim, never fades, is also shown in a glass vase, and is gazed on with reverence by the believers. Some unbelievers (some people doubt everything) say fresh roses are put in every day, but they are probably preserved.

It is the common belief that anything asked of the Virgin of Guadalupe is granted, t have seen people pray with their hands outstretched, and after awhile murmur, "Gracious, gracious!" and get up as if the favor had been received. Women ofttimes kiss the floor when they think they have received mercy at the hands of their dear saint. Near the door are hundreds of rude oil paintings representing scenes in which the Virgin has saved the lives of people. One man fell from a second-story window, and by murmuring the Virgin's name escaped uninjured. Another was not crushed to death, although his horse fell on him. One was released from prison, many from fatal sicknesses, and hundreds of canes and crutches in the corner testify to the many who have been healed.

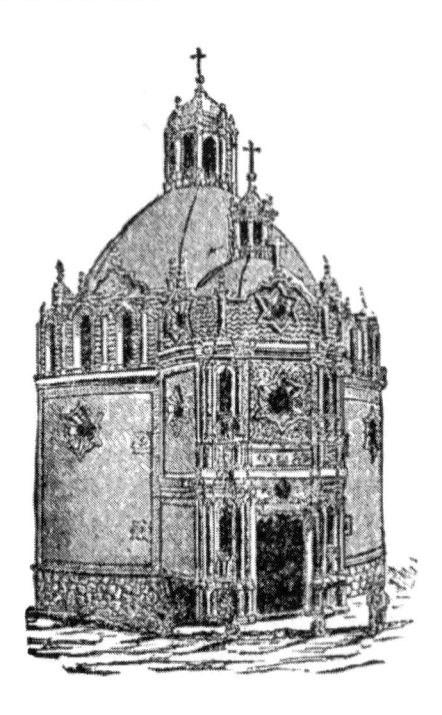

A little green plaza filled with tall trees, beautiful flowers, and flowing fountains, separates the church of the Virgin of Guadalupe from another, which, in order to have some attraction, boasts of a well in the vestibule, which is ever boiling up its muddy water. The water cures any disease, so they say, and at any time a crowd is found around its magic brim filling jars, bottles, and pitchers to take home, or supping from the copper bowl that is chained to the iron bars that cover the well. Very few can suppress the look of disgust when they try to swallow the vile stuff with the all-healing qualities.

Nor are these all the churches of Guadalupe. Away up on the top of a pile of rocks, some hundred feet in height, is the oldest church of the three. It is quite small, and filled with quaint paintings.

At the back of it is the graveyard, where lies the body of Santa Anna, and looking down over the brow of the hill the tourist can see the building where the treaty of peace was signed with the Americans in 1848. It is now used as the barracks. At one side of the church is one of the queer monuments raised in honor of the Virgin, The Escandon family, who are believed to be worth some $20,000,000, once had a vessel out to sea, the loss of which would have put them in bankruptcy. There were great storms, and the vessel had been overdue so long that everybody gave it up for lost. The Escandons went to the church in a body and prayed to the Virgin to restore their property, and they would in return build in her honor a stone sail. It must have been considered a big inducement, for a few days after the ship came in safe, and the stone sail stands to-day a memento of the Virgin's goodness.

Down on the other side, almost at the foot of the hill, is a grotto which, perhaps, is the only one of the kind in the world. A poor Indian formed the rough side of the stone hill into arches, benches, cunning little summer houses and all sorts of retreats. This alone would not have been very attractive, so he came to town and gathered up all the pieces of china, glassware, etc., and, with a cement he had invented, covered every inch with this stuff, fitting them neatly, smoothly and evenly together. All sorts of designs he made – the Mexican coat of arms, pea-fowls, serpents, birds, animals, scenes from life, Eve plucking an apple in the Garden of Eden and handing it to Adam. The work was done so well that it now looks like the finest mosaic, and hence it is called the Mosaic Grotto. Flowers, trees and vines are growing inside, and by candle light it looks like a transformation scene.

There are potteries located here where the Indians make all sorts of queer little things, which have some claim to beauty, and are bought by the natives as well as foreigners. There is some talk of making a pleasure resort at the village of Papotla, the historic Noche Triste, where Cortes, when flying from the furious Aztecs, ordered a short halt, and, sitting down under an old knotted and gnarled cypress tree, wept at his failure. The tree is not a thing of beauty and has very little life remaining in it now; the top has been removed, and it has been badly burned on the inside by someone who had no love for the memory of Cortes. A large iron fence now surrounds it, and effectually blocks the destroyers or trophy gatherer's hand from further vandalism. A pleasure resort might do well here, as the surrounding country is beautiful. Between here and the city is the canal over which the Spanish commander, Alvarado, made his famous leap, thereby saving his life. Stories of it differ.

One says that a wet, mossy log crossed the canal, and the Spanish, seeing this their only means of escape, tried to cross. The condition of the log caused them to slip, and they were drowned in the depths below. When Alvarado came to it and saw the fate of the others, he stuck his spear, or halberd, into the center and safely sprung over. Still others claim he made the leap without the aid of an intervening log.

Another pretty story has been exploded. In the botanical garden at the palace they have the celebrated flower Tzapalilqui-Xochitl, of the Aztecs. The story runs that there are only three of the kind in the world – one at the palace, another at a different point in Mexico, and the mother plant on the mountain. At one time two tribes had a long and bloody war for the possession of it, so the story goes, but with a great deal more exaggeration. The plant is commonly called the "flower-hand," as they claim that inside is a perfect baby hand. I went to see it, and was much disappointed. The tree grows to a good height. The leaves, heart-shape, are thick and about the color of the under part of a silver-maple leaf, except that they are very rough, which prevents them from glistening like the maple. The thick, wax-like, bell-shaped red blossom grows mouth upward, and inside is the so-called hand. It has five fingers and one thumb, but looks exactly like a bird's claw, and not like a hand. The story ran that there are but three in existence. Without doubt the plant is rare and there may be no more than a dozen, if that many, in the world; but I have seen in the gardens of two different gentlemen the very same tree. One of these gentlemen is in Europe, and the other bought his plant from him, so there was no way of learning where the tree came from.

Mexican houses are built to last centuries. It is a common thing to see houses two hundred years old, and they are better than many they are putting up today, for they are adopting the American style of building in as small a space as possible, the structures to stand for a few years. The house where Humboldt lived is near the center of the city. It is not kept as a monument to his memory, as one would suppose when they think of the professed love of Mexico for him, but is occupied by a private family. The only thing that marks the house from those surrounding it is a small plate above the door, on which is inscribed: "To the memory of Alexander Humboldt, who lived in this house in the year 1808. In the centennial anniversary of his birth. The German residenters. September 14, 1869."

At Tacubaya, two miles from the city, there is a large tree, about one hundred and seventy feet in height. It is green, winter and summer, and was never known to shed its leaves, which are of a peculiar oblong shape and a beautiful livid green. For the reason that it never sheds its leaves it derived the name of "the blessed tree;" the large fountain at the foot, which furnishes the water for the poor of the village, is called "the fountain of the blessed tree," and the pulque shop and grocery store opposite are named "the pulque shop and the beautiful store of the blessed tree."

Mexico is the hotbed of children; the land is flooded with them, and a small family is a thing unknown; they greet you at every window, at every corner, on every woman's back; they fill the carriages and the plaza: they are like a swarm of bees around a honeysuckle – one on every tiny flower and hundreds waiting for their chance. A man died the other day who was followed to the grave by eighty-seven sons and daughters, and had buried thirteen, more than you can count in three generations in the States, so he was a father to the grand total of one hundred children. There is another man living in Mexico who has had two wives, and who has living forty-five children. Down in a small village, out from Vera Cruz, is a father with sixty-eight children. Allowing the small average of five to a family, one can see how numerous the grandchildren would be. I am acquainted with a gentleman whose mother is but thirteen-and-a half years older than he, and she has eighteen more of a family. It is a blessed thing that the natives are able to live in a cane hut and exist on beans and rice, else the lists of deaths by starvation would be something dreadful.

CHAPTER XIX.
A DAY'S TRIP ON A STREET CAR.

AFTER being annoyed by the porter for two hours, who feared we would miss the train, our party of two at four o'clock in the morning started for Jalapa. Even at this unholy hour a large crowd had gathered at the station, where they busied themselves packing their luggage aboard. Every woman had one male escort, some several. The Mexicans surveyed myself and my chaperone in amazement, but I defied their gaze and showed them that a free American girl can accommodate herself to circumstances without the aid of a man. The mozo who had carried the bothersome sachel demanded "un peso" (one dollar), which I very promptly refused, and gave him the smallest change from my purse – twenty-five cents. The seats run lengthwise, like in an ordinary street car, and a Frenchman sitting opposite, who witnessed our little transaction and my very limited knowledge of Spanish, remarked: "Well, mademoiselle, you are smarter than I. A man charged me one dollar and a half just for the same service that one rendered you, and, although I speak Spanish, I had to pay it."

The occupants of the car were the Frenchman and his wife, a musician, wife and sister-in-law, a Mexican and Frenchman solitaire, as they say here, and ourselves. It was far from daylight, so, making themselves as comfortable as possible, they all went to sleep. The Mexican women were dressed in plain black, with black veils and very high hats; they carried little black hand sachels, wore no gloves, and their finger-nails, easily a half-inch longer than the finger, were cut in the bird-claw shape then so fashionable. The Frenchwoman did not look very pretty, as she slept with her mouth open. She was dressed in red silk, with red hat and veil, yellow gloves and linen duster. She was very fleshy, and had, besides a hand sachel, a cage in which were two brown birds dotted with red, which they informed us later were French canaries. Her husband was about six feet three inches, and weighed undoubtedly three hundred pounds. The solitaire Frenchman was bald-headed, and had white side-whiskers, which stood out at right angles to the length of one foot; his whiskers were the largest part of him. The Mexican had a very red nose, extremely thick lips, and was rather effeminate-looking. The married Mexican looked exactly like a jolly Irishman – something very extraordinary. After I had finished this inspection by the dim light of a lamp which hung in the center of the car, I too went to sleep, and knew no more till the train stopped at the journey's end, a few miles out from Vera Cruz.

It ended the train's journey, but not ours, for the rest of the trip is made by tramway. The cars are very high, have four seats, and the rays of the sun are excluded by a tin roof and canvas sides. Six mules do the hauling, and two cars – first and second-class – are run each way daily. They run on a regular iron track, as it was once the intention to run steam cars here. A great deal of freight is hauled in this manner. The village surrounding this station is entirely composed of straw huts. We were soon seated in the tram car, our number increased by the guardsmen, who, as the old saying goes, were armed to the teeth. A bell rang, and off we started with a rush, the second-class car keeping close to us. Our happiness would have been supreme had not the driver lashed his mules continually. The scenery was fine. The tall, graceful palms, the cocoa trees, the thousands upon thousands of beautiful orchids and wild flowers, the many-colored birds, some piping heavenly strains, others taking their morning bath in the running stream which crept along the wayside with a dreamy murmur; the delightfully fragrant, balmy air, everything seemed to lend its aid to make the scene one of indescribable loveliness. It was interesting to note the homes and home life of the natives in this rural spot; their straw houses are built simply by setting trees for corner posts and sticking the cane into the ground around them. The roof, of cane, grass, or palm leaves, always runs up to a high peak. Generally, every house has a porch and more rooms than one, but never any other floor than the ground. Sometimes they exhibit good taste in building and one house will have several rooms, two or more porches and pretty peaks and curves which one would think impossible to make of cane; the furniture does not cost much, it consists entirely of petates; they furnish the tables, the beds, the chairs, and, suspended by a rope, make a comfortable swinging cradle for the babies. This useful piece of furniture is nothing more or less than a mat, woven by themselves in plain or colored straw; these people, no difference how poor, own burros, dogs, chickens, pigs, and other domestic animals, which do not occupy outside or separate houses, but live, sleep, and eat right in among them. A pig is as much at home in the kitchen or parlor as in a mud puddle. It is no uncommon sight to see sleeping children bound on one side by a pig, on the other by a sheep, and at their feet either a dog or a goat.

Dinner was secured at an inn situated midway on the line. The landlord taxed each passenger one dollar for the frugal repast, and even then, did not seem satisfied. The rays of the sun were beating fiercely down when the travelers again boarded the tram car. One woman took from her sachel a cross and prayer book, and read herself to sleep. The other Mexican girl leaned her head on the back of the seat and went to sleep. The big Frenchwoman turned her back to the side of the car and putting her knees up on the seat she, too, went to sleep. Her husband by this time was nodding slowly and soothingly, while the other Frenchman was trying to tickle him by running a straw down his back, but at length he tired of efforts unrewarded and sat down and went to sleep. When I looked at the two Mexicans they were asleep, one with a half-smoked cigarette in his mouth. The driver had tied the lines around the brake lock and was in the midst of the land of nod. Even the two holders of defensive weapons, who were there to guard us from all sorts of imaginary evils, were so sound asleep that a cannon shot would not wake them. Even the little birds had tucked their heads carefully under their wings and, maybe, were dreaming. It was all so comical that I glanced at my little mother to find she was bravely trying to resist the sleepy god. She gave me a drowsy, sympathetic smile, while I buried my face in my light shawl and laughed just like I used to do in church when I would see anything funny, and my laughter was just as hearty and hard to control. The mules had long ago gone to sleep, but still managed to move slightly. The situation was too overpowering, and I must confess that after putting myself into as small a knot as possible I deposited my entire body on the seat and soon went sound asleep.

When I opened my eyes, I found all the rest awake and the married Mexican preparing to shoot birds. The driver was certainly the most obliging fellow in the world. When anything was shot he stopped the car and waited until the other got off and procured his game. The Mexican shot at everything which was living, except the trees and flowers, but he got off for nothing but squirrels, and the heartlessness of it made us wish they had a humane society here, for many of the poor birds were disabled, and the thought that they must live on in pain for many days was not a pleasant one.

Our route lay over the old diligence road that connected Mexico with the end of the world. Cortes, the French and the Americans all traveled over it. We crossed the old national bridge and saw the ruins of one of the forts built by Cortes. When the Mexican tired of his killing sport the three ladies joined him in a game of cards, which the passengers and driver watched with absorbing interest, while the mules resumed their nap. I was bored beyond endurance by the listlessness of the company, and was not sorry when their attention was attracted by a cart drawn by four oxen, which was descending a high hill in the distance.

The cards were put aside, and they began to talk about the hacienda, which was clearly in view, and the beautiful mansion, cathedral, and numerous homes for the laborers, which held a commanding position on top of the same high hill down which the cart was coming. When we reached the brow of the hill, by looking back, we could see a white streak which separated sky and earth, and were told it was the sea at Vera Cruz, sixty miles away. The cart stopped at this point, where the motive power was renewed by fresh mules, and its passengers – three women – kissed and hugged the trio of Mexicans in our party. The hacienda, owned by our fellow travelers, once belonged to Santa Anna. When we resumed the journey, it was drawing on toward evening, and I began to view the beautiful surroundings with but a lazy interest; the queer fences, built of mud and topped with cactus plant, and hedges formed of beautiful palms, fifty or one hundred years old, commanded but a passing glance. Pretty little homes, lovely gardens and sugar factories had ceased to be of interest, so we settled down to rest until the Frenchman stretched out his arm and ejaculated "Jalapa!"

In a moment all weariness vanished, and we were fresh as in the morning. I wish I could show you Jalapa just as I saw it then. It nestled down in the valley like a kitten in a cushioned basket. The white houses gleamed like silver through the green trees, while the surrounding mountains were enveloped in a light bluish mist which grew black as the distance increased. The sun had just slipped behind one, leaving its golden trail, the black and white clouds, the misty mountains all mixed in one harmonious mass. We entered the town with a rush, the driver blowing his tin horn to warn the inhabitants of our arrival. A large crowd had collected at the station, but only two hotel runners were there to bother us, and as all the other passengers were citizens they clung to us faithfully. The Frenchman said he would go with us to the hotel and make all arrangements. He took us to what he thought was the best, and asked the woman the price. "One dollar and fifty cents a day," she said, and as we were satisfied he bade us good-bye, and left us to the tender mercies of the Mexicans. The hotel was certainly very clean and nice. In the courtyard were trees and flowers. A porch paved with brick tile surrounded this, and was hung at every available space with bird cages. The building, only one-story, was painted white, with trimmings of blue. The overhanging roof was down low, and the rafters, which are never hidden, were painted a light blue. The supper was undoubtedly the best we had eaten in Mexico, and it immediately put a warm place in our heart for the little superintendent, who lived awhile in the States and there learned to cook.

Jalapa is at present the capital of the State of Vera Cruz; the capital business is very different here from what it is in the States; there, once a capital, always a capital; here, every new Governor locates the capital where it best suits his convenience, if that should be in the forest. Orizaba and Vera Cruz have both served repeated terms, but Jalapa made a successful run and got in at the last convention. It is a very old town, and not only noted for the beauty of its women, who possess light hair and eyes, and beautiful complexions, but for the beauty of its location. It is known as the flower garden of Mexico, and the old familiar saying was, "See Jalapa and die," as it was supposed to contain everything worth seeing; but at present it is simply a beautiful, sleepy paradise, reminding one of a pretty child in death – quiet and still, almost buried in a wealth of flowers; the government buildings and churches are very fine, but the houses are only one story; they are built with low, red-tiled, overhanging roofs, and are tastily painted. Some pink houses have light-blue overhangers and *vice versa*, while white houses have blue or pink, and the yellow have blue, pink, and white trimmings. Every street is very irregular, narrow in some places, wide in others, and as crooked as the path of a sinner. One can walk for a day and imagine they are on the same street all the time, or on a different one every thirty feet, just as fancy dictates.

One would willingly spend a lifetime on this "spot of earth let down from heaven," as the Mexicans speak of it. Away over hills and ravines can be seen the great Cofre de Perote, thirteen thousand five hundred and fifty-two feet high. A great mass of white porphyry, in the shape of a chest, gleams from its dark side. From this it derived its name, "Cofre." Still above all, as though endeavoring to reach heaven, is the snowy peak of Orizaba. The former is within a day's travel from the town, and well deserves a visit. To the northeast, thirty miles distant, is the lovely village of Misantla, noted for its beautiful scenery and Aztec temple and pyramid. A little further north is another pyramid, the finest and oldest in Mexico. Jilatepec is only seven miles away. It is a lovely Indian village, peculiarly situated at the bottom of a deep valley. Several foreign families are located at the flower town of Cuatepec, owners of some of its far-famed coffee plantations.

Jalapa has a population of 12,400, and an elevation of 4,335 feet. The climate is cool, the soil fertile, and the town never visited by contagious diseases. All around are plantations of coffee, tobacco, vanilla, cotton, maize, and jalapa – the well-known old medicine which was a remedy for every known ill to which flesh is heir to. Jalapa is pronounced as though it were spelled with an *h*, with a soft sound to the *a* – Halapah. There are many cotton mills around the suburbs that are well worth the time it takes to visit them. We visited one owned by our polite French friend. The building once sheltered nuns, and the garden which surrounds it shows what it might once have been, but is now one tangled mass of climbing roses, lemon, orange and coffee trees, and numerous flowers for which I know no name. At the back, from a little stone turret, one can view a smooth green plain divided by a silvery stream – known by the inappropriate name of the Dry River, while it was never known to go dry – which flows on to make that ponderous machinery its slave, as it turns around with almost diabolical glee. Men and women do the work. They receive from one real to seventy-five cents a day. The machinery all comes from England.

Not far from the main Cathedral are the ruins of the Convent of San Francisco. It is easily three hundred years old, and is of immense size. Over the door of the chapel part we could trace "Property of King Philip, of Spain," while cut in gilt letters on a black plate, just a little nearer the edge of the building, is the inscription, "Land of Benito Juarez." The baths are now used for the benefit of the public, costing only six cents. The open swimming baths are used for horses and dogs, the former costing three cents, the latter gratis, providing the canine accompanies the horse. The public laundry is another place of interest. It is situated in the center of the town, built of brick, with stationary porous stones for washboards. The city charges nothing for the use of the place.

When evening came I called my old landlady up and offered her three dollars for the day. "No," she said, "I want six dollars." I was astonished, but managed – with a mixture of English and Spanish – to tell her I would pay no more. She went to her husband and he made out a bill "payable by Nellie Bly for two – supper, all night, coffee and breakfast, six dollars." I told her it was all wrong, and added that she was bad, because I did not know a Spanish word for cheat, but I wanted to get as near it as possible. At last I tried to drive some sense into her head, and explained that the bill for one day for two was three dollars." "Si hay" (pronounced 'see eye"), she asserted. "Well, I came last night, was here till this afternoon; one day, eh?" "No, two," was her astonishing reply. "Well, madame, twenty-four hours is one day in the United States, and if it isn't so here, I will start it now." I gave her three dollars; and, remembering the old adage that "he who fights and runs away lives to light another day," and having no desire to leave my bones in Jalapa or go to Vera Cruz with a map drawn on my face with her finger nails, returned to my room and left her to vent her rage on her husband or servants, as she wished. But she was not going to be beaten by a "gringo," so she sent for the Frenchman who brought me there. He rapped on my door, and asked what was wrong. I told him the old lady was not only seeing double, but counted everything by the second multiplication table. He laughed, and said she thought I was a "gringo," and she could cheat me. He soon made her see clearer, and we remained the following night and had supper for seventy-five cents. I had learned pretty well how to make all arrangements first, and proposed in the future not to drink a glass of water until I knew the price. I had no intention of allowing a Yankee girl to be cheated by a Mexican, man or woman.

The next morning, we started on our return trip to Vera Cruz. We looked forward to it with pleasure, as the former day spent on a street car was one of the most pleasant and unique experiences of my life. We had very few passengers down, the conductor, two soldiers, driver, one old woman and ourselves, and a game rooster, who crowed at every village, and was treated with as much consideration as a babe would have been. At the station, just before we started, an old man who had heard us speaking English, came up and spoke to us. He was an American, but having lived in this town for forty years had forgotten his mother tongue. His English was about as good as the newsboy's who took me to his hotel in Vera Cruz. The old woman was going about one hundred and twenty-five miles to see her married daughter, and she was bare-headed. This woman did not know there was such a thing as the United States, could not imagine what New York meant, and had never heard of George Washington, not to mention the little hatchet and the democratic cry of "If at first you don't succeed, try, try again." She made the day's trip alternately smoking a cigarette and reading her prayer-book. A short way out on the road the driver got off and picked up a little gray bird by the roadside. On examination. I found its side was terribly lacerated by a shot, but I bound it up with my silk handkerchief and decided to carry it to Vera Cruz, where I would try my hand at surgery. The day passed similar to the former one, everybody going to sleep after dinner; but the beauty of the country, and the novelty of a day in a street car, robbed it of all disagreeable features, and as we neared Vera Cruz I not only noted this the spiciest experience of my life, but said I would not exchange it for any other in the Republic of Mexico.

CHAPTER XX.
WHERE MAXIMILIAN'S AMERICAN COLONY LIVED.

ON opening my door one morning to leave for the railway station a man, who had evidently been waiting by the side of the entrance, sprung forward and seized my baggage. My first impression was that he was a robber; but I retained my screams for another occasion and decided it was a mozo who wanted to help me to the train. Remembering former experience, and wishing to profit thereby, I rushed after and caught him just at the head of the stairway. Clutching his blouse with a death grip, I yelled, "Cuanto?" "Un peso," he answered. Well, as I was a healthy American girl, and as strong as one can be after several months' training on beans and cayenne pepper, I had no intention of giving a great, big, brown fellow $1 for carrying a five-pound sachel half a square. I said "no" in a pretty forcible manner, and gave weight and meaning to my monosyllable by jerking the sachel away. He looked at me in amazement, and as he saw I was not going to be cheated he said fifty cents. I said nothing, and, picking up the sachel, trudged downstairs. At the door he once more approached me and asked how much I would give. "Un medio" (six and a quarter cents), I replied. "Bueno," he said, and took it at the price, while I congratulated myself on saving ninety-three and three-quarter cents.

The car was full of people who, we found out afterward, composed a Spanish opera troupe. Although they were not many they filled the car, and in order to get a seat we had to put down shawls, beer and wine-bottles, band-boxes, lunch-baskets, a pet dog, a green parrot, and numerous small things. Every woman had at least three children, which were cared for by as many nurses. Oh, what a howling, dirty, lazy mob!

The pretty little town of Cordoba lies about two miles from the station, and street cars, hauled by four mules, await each train and carry the passengers to the village – first-class, twelve and a half cents; the cars wind through little streets shaded on either side by beautiful foliage, which, every here and there, gives the tourist tantalizing glimpses of the exquisite tropical gardens within; the street car passes the only hotel in the town – the Diligencia. It is a low, one-story structure, and looks more like a cattle-yard than a habitation for human beings; the overhanging roof droops toward the pavement, and is within a few feet of the ground. Inside one sees a little porch on one side, which, covered with many trailing, curling vines, serves for the dining-room. Opposite is an office and bedroom combined, where, at the desk, sits a grizzly-haired man writing, ever writing, from morning until night's shade hides the tracing from his aged eyes.

He greets one with a weary, pathetic smile, and a faraway look in his saddened eyes, as though wondering what has become of all the guests who used to trip in gayly, with black eyes and white teeth sparkling in evident pleasure at reaching his hospitable board, with whom he grasped hand, and in true Mexican style said, "My house is yours," and that friend responded, "Your humble servant." Poor old landlord, he has lived too long! The advent of civilization has rushed in upon his friends and crushed out his trade. The noisy old diligencia has long ceased to rattle except in his memory, and the modern street-car stops at his door once in many months to leave him a white-faced, curious stranger, whom he greets with that strange smile and then returns to his writing, waiting for that which is nevermore.

A man and woman came in on the same train, and the latter offered her services to us, being able to speak the two languages. When we entered, the chambermaid took my troublesome baggage and led us back to where the rooms formed a circle around the court. In the center stood a large basin where several old horses and mules – which looked like old "Rip" after his long sleep – were lazily drinking. They paused long enough to survey the unusual arrival. When we entered our room the chambermaid – who is always of the male gender in Mexico – set down my baggage and demanded fifty cents. I, not feeling disposed to throw money away, decided not to pay one cent. Accordingly, I laid aside my few words of Spanish and spoke to him in English. "What do you want? I don't understand," etc. At last he took two quarters from his pockets and held them before me on his open palm. I calmly reached out, and, taking them, was going to transfer them to my pocket when he, in great alarm, yelled: "No, no!" and grabbing them, tied them up in the corner of his handkerchief, with great haste and evident pleasure. It had the effect of curing him, for he immediately shook hands and left without demanding more.

Cordoba, or Cordora, was established April 26, 1617, with 17 inhabitants. It was during the time of the Viceroy Diego Fernandez de Cordoba, Marquis of Guadalcazar, and was named for him. King Philip III. of Spain issued the charter on November 29 of the same year. The population to-day, composed of Mexicans, 2 Germans and 1 American, is 44,000. It is built compactly. The town is clean and healthful. Nearly all the streets are paved, but everything has a quiet, Sunday-afternoon appearance. There are no public works, but the surrounding plantations, which mark it as one of the prettiest places in Mexico, furnish work for the populace. The Indians are cleaner and better looking than those around the City of Mexico, and children are not so plentiful. But one pulque shop is running, consequently there are less drunken people than elsewhere, yet the jail is full of prisoners. On Sunday people are permitted to visit their friends in jail. They cannot go in, but they can go as far as the bars and look through. The prisoners are herded like so many cattle. Their friends carry them food. They push a small basket through the bars, and the intervening officer puts it through another set of bars into the hands of the fortunate receiver. Sometimes the prisoners get a few pence and are enabled to buy what they want from the venders who come there to sell. Indeed, it is ofttimes difficult to say which mob looks the worse, the one on the inside or the visitors.

The market at present is situated on the ground around the plaza, but some well-disposed Spanish gentleman is building what will be one of the handsomest market houses in Mexico. It is situated on the edge of town, and the surroundings are most pleasing. On one side is the ruins of an old convent, famous for the goodness of the sisters, their exquisite needlework, their intelligence and beauty. But time has laid his hand heavily on the structure, and it has fallen into decay. At the back stands a high marble shaft, broken at the top, and dotted with green cacti which have sprung forth from the little crevices. It has the appearance of very old age, but was erected in honor of those who fell in the fight for liberty. One of the finest gardens in Mexico bounds the other side. It is the property of the gentleman who gave the ground and is building the market house, which alone will cost $50,000. It is a marvel of beautiful walks and cunning retreats. It seems absurd that such a spot, so fitted for love-making, should be placed in a country where they don't know how to make use of it. In the center stands a Swiss cottage built of cane, with a stained-glass window.

A stairway, also of cane, leads to the second story, and little balconies surrounding the colored windows give one a lovely view of the entire valley and surrounding hills. I wish it were in my power to give some idea of the bountiful flowers which are forever opening up their pretty perfumed faces in this entrancing spot; there are thousands of roses, of all colors and shades, from the size of a gold dollar to that of the fashionable female's hat. One spot shows tiny flowers fit for the fairies, of wonderful shade and mold; next would be a large, healthy, rugged tree, which bore flowers as delicate and dainty as any plant in existence. It reminded one of a strong father with his tiny babe in his protecting arms; the handsome avenues are perfect bowers of beauty; the little birds in the foliage twitter softly but incessantly. It is all life, but in a subdued, gentle monotone, soft as the last lullaby over the little child who has closed its eyes and, with a smile, joined that heavenly band to which it rightly belongs.

This is the only place in Mexico where we found a man who knew enough to have the flowers separated by a green lawn. It is the universal rule here to grow anything but grass, which is considered an unsightly weed. A Spanish gentleman once took me to see the grounds surrounding a Mexican mansion. The trees, flowers, and shrubs, as well as the statuary and fountains, could not be excelled, but the ground was bare as Mother Hubbard's cupboard, and swept as clean as a dancing floor. "This place cost more than five million dollars, and thousands more yearly," explained the gentleman. "You have nothing in the States to compare with it."

Cordoba supports three public schools and male and female academies, one theater and about thirty churches. The finest church, located next to the plaza, cost thousands of dollars. It has a marble floor and twenty altars, dressed in the finest lace, with silver and gold ornaments. The frescoing displays exquisite workmanship. The images are wax-clad, and quaint.

The plantations surrounding Cordoba grow oranges, pine apples, coffee, bananas, tobacco, rice, cocoanuts and peanuts. Coffee was introduced into the West Indies in 1714, and here in 1800. It grows best in a temperate zone, and Vera Cruz raises more than any other state in Mexico. Most every variety requires protection from the sun, and will die if set out alone, so those having large groves plant coffee in them. Others make double use of their fertile land by planting groves of cocoa palms with the alternate rows of coffee trees. The leaf and bark of a coffee tree resemble that of a black cherry. The blossom is white and wax-like, with a faint perfume, and the berries grow on a branch like gooseberries. A tree will bear three years after planting the seed, and on one branch will have ripe and green coffee and blossoms all at the same time. When ripe it is gathered and laid on the ground to dry, being stirred every morning to dry it equally. This whips the hull off, and it is taken to the village, where it sells for four cents a pound. Each hull holds two grains. One tree will live and bear, with little or no cultivation, for eighty years. Bananas are four years old before they bear. The finer banana is never seen in the States, as it will not bear shipping. The kind shipped there the people here consider unfit to eat unless cooked, and they prepare some very dainty dishes from them. There are more than fifty different varieties, from three inches in length to three-quarters of a yard. The small ones are the best. The leaves are used by the merchants for wrapping-paper, and by the Indians for thousands of different things.

Tobacco now grows in about half the states of the republic, and thrives up to an elevation of six thousand feet. Formerly its cultivation was restricted to Orizaba and Cordoba, and a leaf of it found growing elsewhere, either accidentally or for private consumption, was, by law, promptly uprooted by officials appointed to watch for it. In 1820 two million pounds of it grew in this district, but now the output is greatly decreased, owing to the heavy taxes. Sugar cane grows in all but six states, up to an elevation of six thousand feet. It requires eighteen months for crops to mature, except in warmer soil, when it takes from eight to ten months.

One remarkable thing is, that the men who own the fine gardens surrounding the village do not live near them, as one would suppose, but inhabit stuffy little houses in the midst of the town. One bachelor has on his plantation plants from all parts of the world, over which he has traveled ten times. He cultivates all kinds of palms in existence, among which we noticed what is known as the "Traveler's tree." It is a strange looking thing, with long, flat, thick leaves growing up as though planted in the center and hanging loose at the ends. The flower is beautiful, with three long petals, the upper two white and the under one a sky blue. It is of a wax-like stiffness. Readers of books of travel will be familiar with the tree, it derives its name from the fact that it grows in the desert where no water is to be found. On thrusting a penknife into its body, a clear stream of water, probably a pint and a half, will flow from one cut, and people traveling through the desert quench their thirst from this source, hence its name. The water is very cool and has a slight mineral taste, but is rather good and pleasing. It gives water freely all day, but, after the sun sets, is perfectly dry.

The bread and quinine trees are among his interesting collection. One odd plant attracted attention. It bore a round, green leaf, but wherever there is to be a blossom the four leaves turn a pretty red and form a handsome flower, each leaf forming a petal. The true blossom, which does not amount to much, being long and slim, like a honeysuckle, forms the stamens. It is of foreign importation, and grows in a climbing vine, whole arbors being covered with it. The grounds are surrounded by a hedge of cactus, which is strong and impassable. The Yucca palm and fruit cactus grew off in a corner by themselves. Several small streams run through this plantation, spanned by lovely rustic bridges. In the deep ravines are found ferns of every variety known, and on the trees a collection of orchids which, I believe, has no equal in any country. The happy owner, who is a bachelor worth about $20,000,000, lives in a little house in the center of this town, which has never been furnished until last winter, but in the courtyard, he has plants from every country in the world, for which the shipment alone cost $40,000.

Down by Cordoba I found a tribe of Indians who are not known to many Mexicans excepting those in their vicinity; they are called the Amatecos, and their village, which lies three miles from Cordoba, is called Amatlan; their houses, although small, are finer and handsomer than any in the republic. Flowers, fruit, and vegetables are cultivated by them, and all the pineapples, for which Cordoba is famous, come from their plantations; they weave all their own clothing, and have their own priest, church, and school. Everything is a model of cleanliness, and throughout the entire village not one thing can be found out of place; the women are about the medium height, with slim but shapely bodies; their hands and feet are very small, and their faces of a beautiful Grecian shape; their eyes are magnificent, and their hair long and silky; they dress in full skirt, with an overdress made like that we see in pictures of Chinese women, or like vestments worn by priests of the Catholic Church. It is constructed of cotton in the style and pattern of lace. Around the neck and ends it is beautifully embroidered in colored silk, the dresses always being white. On the feet they wear woven slippers of a pink color, and on their heads a square pink cloth long enough in the back to cover the neck, like those worn by peasant girls in comic operas; the arms are bare, covered alone with bands and ornaments; the neck is encircled with beads of all descriptions, and is also hung with silver and gold ornaments; the ear-rings are very large hoops, like those introduced into the States last fall; they never carry a baby like other tribes, but all the children are left religiously at home.

The men are large and strongly built, not bad-featured, and wear very white, low-necked blouses and pantaloons, which come down one-third the distance between waist and knee. They also wear many chains, ornaments, bracelets, and earrings. They are always spotlessly clean, and if they have a scratch on their body – of which they get many traveling the thorny roads – they do not go outside their village until entirely healed. They are industrious and rich, and never leave their homes but once a week, where they bring their marketing and sell to the Indians in Cordoba, as they are never venders themselves, selling always by the wholesale. Their language is different from all the others, but they also speak Spanish. The women are sweet and innocent. They look at one with a smile as frank as a good-humored baby's, and are undoubtedly the handsomest and cleanest people in the republic. I would not have missed them for anything, and can now believe there are some Indiana like the writers of old painted them.

In the time of Maximilian, a colony of Americans asked the emperor for land on which to settle. He kindly gave them their own choice, and they settled at Cordoba, where they had the advantage of the tropical clime and were secure from yellow fever. They were three hundred in number, and in a short time, with true American industry, they made business brisk. Three American hotels were established, and the plantations were the finest and most prosperous in the land. Maximilian looked on the little band with favor and gave them ample aid and protection. During the rebellion the liberty party made raids on their homes, destroyed their property, and not only made them prisoners and hurried them off to Yucatan – a place from which there is no escape – but murdered them whenever they wanted some new amusement. Maximilian was powerless to help those who had prospered under his care, and just when he was to be shot the last of the colony, who feared the liberty party, deserted their once happy homes and went to another country. Only one remained, Dr. A. A. Russell, who has been the solitary American here for twenty years. The hotels have disappeared, and the plantations, now possessed by Mexicans, bear no traces of their once tidy and prosperous appearance; this is the history of the first and last American colony ever formed in Mexico, given me by the last remaining colonist, who reminds one of the last chief, inconsolable and disconsolate, keeping vigil at the tombs of his people until death shall claim him too.

CHAPTER XXI.
A MEXICAN ARCADIA.

"IF you come over here you will get a better view," spoke a gentleman as he came from the back end of the car hauling us from Cordoba to this place. We were nearly breaking our backs in a vain endeavor to look over a man and wife, surrounded by almost as many children as belonged to the old woman in the shoe, down the perpendicular side of the mountain into the deep ravine beneath. We took a survey of the speaker, of his light woolen suit with wide sombrero to match, his pleasant, handsome face, and mentally decided that he was not only worth looking at, but also worth talking to. By the time the train had passed the barranca we were in a deep conversation, quite after the manner of Americans, and although none of us asked any impudent questions we were discussing marriage and women's rights.

"I think every woman should be taught some useful occupation," he said, "and their education should be unlimited. But the one great fault of the world is not paying a woman what she is worth. There are few things in which a woman is able to sell her talents at the same price as a man, and it is a reproach to humanity that it is so. I have three daughters now at school. The eldest is studying to be a physician, the second has great artistic ability which she is cultivating, and the third is a good musician. In either of these vocations they can take their place among men and receive the same recompense.

"I am living in Orizaba now," he continued, "and have been hunting deer for the past few days just below Cordoba. We saw plenty, but our man and dogs did not understand the game, so we returned empty handed. The only thing wounded is my friend back there, who fell out of a hammock while we were away and sprained his ankle." As we told him Orizaba was also our destination, the next question was where did we intend to stop, and found it was the place where he lived. After he had given the wounded man into the care of friends, we got on a car and soon reached our hotel. It was so dainty and nice that I cannot resist a brief description for the benefit of those who may someday be in its locality.

It is known as the La Borda, and is near the station, as well as the best in the town. The rooms are a model of cleanliness and neatly furnished. From the front one can survey part of the village, and the range of mountains outlined against the sky like immense waves, each one climbing higher, and above all the great mountain, that majestic monument which wears its snowy nightcap seventeen thousand two hundred feet above the level of the sea. At the rear of the house, just below the dining-room windows, is a never-ceasing waterfall which goes to feed some mills in the vicinity. In the first glimmer of day with our wakening senses we hear its murmuring song with that of the birds. Its sound is in a gentle, half-subdued manner, as though enticing the birds to come nearer to its brink and bathe their toes and quench their thirst with its foaming waves. Near midday it gets loud and boisterous, and you seem to hear: "The day is short, improve your time," over and over with a monotony that rather fascinates us.

Directly above this wonderful fall is a cozy little garden, cultivated by the landlady, who also deserves a word. She is a German, who accompanied her husband to this country some years ago. He died and left her in a strange land with two baby girls, whom she maintains by running this hotel. She is quite pretty, and speaks German and Spanish fluently, while she is studying English, and understands some now. She keeps her house, like most Germans, as clean as it can possibly be made, and endeavors to have all her guests feel at home. The cooking is so good and everything so comfortable that one would fain have the little German woman and the La Borda in every town in the republic.

Orizaba is a beautiful little valley surrounded by a chain of majestic mountains. The houses are white and most generally of one and two stories. There are 25,500 inhabitants. It was for a long time the capital of Vera Cruz. When this place was first founded in the year 1200 by the Tlascaltacas, its original name was Ahanializapan. which, translated, meant "Pleasure in or on the water." The people prospered and lived in peace and happiness until the Aztec Emperor Montezuma reduced them to his dominion in 1457. Still under such a good and wise king they could not be otherwise than happy in this lovely garden, until Gonzalo Sandoval undertook and was successful in conquering them in 1521. But even war did not stop its progress, and in ten years later, in 1531, the governor gave it its present name, the Valley of Orizaba. The people grew in intelligence, and were industrious and religious. In 1534 they built their first parish church, Gonzalode Olmedo, and as early as 1599 had put up a building and opened their first school. Inhabitants increased rapidly, and in 1774 it took the rank of town. Not satisfied yet, they built up, and the population increased by birth and new settlers until in 1830 it was declared to be a city.

Orizaba was for a long-time capital of its state, Vera Cruz, and is now the pleasure and health resort for people from all over the republic, besides being the home of the wealthy people of Vera Cruz. No yellow fever or any of the other diseases come to this dainty valley, yet twelve doctors are holding forth and trying to gain a living in the vicinity. All are Spanish, with the exception of one, an Austrian, and only two speak English, one of whom used to write for an American paper. For the entire population there are but three baths (banos), but the poor can go to the river which runs nearby. The only amusements are the billiard hall, bowling alley, and two fine theaters. One contains 272 lunetas, eighteen plateas, nineteen palcos, and one galeria. The other cost $100,000, and has a magnificent interior. It has 252 lunetas, eighty balconies, three grilles, thirteen first-class and thirteen second-class palcos, and one galeria.

On the map there are recorded but eleven churches, but even from our hotel window we could count many times the number. Those recorded are the San Antonio, Calverio, Concordid, Las Dolores, Santa Gertrudes, San Jose de Gracia (ex-convent), San Juan de Dios, San Maria, Tercer Orden and La Parroquis, which is the largest and finest. It is situated in the zocalo and has had its steeple knocked off three times by earthquakes. The latter seem to have a special grudge against this one church, for although they have caused the towers of many others to lean, they have never shaken any of them completely down. Orizaba must be a very naughty child – beautiful children most always are – for Dame Nature often gives it a shaking. She is an indulgent and not very severe mother, as little or no damage is ever done by the correction, excepting to this one cathedral. During our stay the earth shivered as though struck with a chill, but the people paid no more attention to it than we do to a summer shower; not half so much, in fact, as we do when the mentioned shower threatens to ruin our Easter bonnet.

Two little Spanish papers of four pages, or two sheets, about 8x6 inches square, retail at twelve and a half cents and furnish the news for the inhabitants. The children here should not be lacking in education, as there are ten schools for boys and six for girls; they can start at any age, and go as long as they wish. Besides this, the government sustains a preparatory college of one hundred and fifty students, at the yearly cost of eleven thousand dollars; a high school for girls, two hundred and fifteen pupils, at two thousand eight hundred dollars, and a model school for boys, one hundred and eighty students, at five thousand six hundred dollars. The government also gives a subsidy to five adult schools of six hundred dollars. The municipality schools, four for boys, three for girls and five for adults, cost yearly eight thousand dollars. In addition, there are twenty-nine private schools, with an attendance of five hundred and forty girls, six hundred and forty boys and sixteen adults; yet, with all this well-made report, there are in the Republic of Mexico two million five hundred thousand people who cannot read or write.

Orizaba has rather a big heart – they furnish a free home for men and one for women with hospitals attached, but one don't dare mention their cleanliness or order; they are under the superintendence of the Board of Charity. There is also a retreat for the insane, which, like ours in the States, occupies a spot free from all other habitations. The last year's report of the town's statistics shows that they received indirect contributions, $25,000; direct contributions, $20,000; miscellaneous sources, $4,000; municipal rights, $4,000; contribution of twenty-five per cent, to Federal district, $27,000. Pulque shops are scarce, there being only three, besides one lithographer, one public garden, two photographers, one dentist, four established cigarette manufactories, and one lottery, for it is impossible to find a Mexican town without. There are no Americans in the town, except those who belong to the railroad.

Many things of interest are to be seen in and around Orizaba. One who cares to climb can ascend the Cerro del Berrego and view the old ruins which mark the spot where the Mexicans were defeated during the French invasion, June 13, 1862. A little way out is Jalapilla, where Maximilian resided a short time after the French army had gone, and where he held the famous council to determine whether he should abdicate or not. One and a half miles south are large sugar plantations and mills. Besides, there are several waterfalls, between two and five miles distant, noted for their beauty and strangeness; the Cascade Rincon Grande is about one mile east; the water has a fall of over fifty feet, and all around is a luxuriant growth of vegetation, which helps to make the spot one of the prettiest in Mexico. Donn Tonardo Cordoba is a forty-foot fall, which disappears in a round hole in the earth, falling to a depth that has never been measured.

Another thing interesting to foreigners are the old Spanish deeds, written on parchment during the time of Cortes. They can be seen at the register's office by giving the man in charge two reals for his trouble. On Sunday afternoon bull-fights are held in an old convent, and what was once a fine church is now the barracks for a garrison and hall for the Masonic lodge.

Many people have a fancy to climb the peak of Orizaba, which is 17,200 feet high. It requires but five hours of a good climb to reach the summit. The last eruptions it had were in 1545 and 1566. Several times it has been reported smoking, but the rumors were finally, on investigation, pronounced unfounded. The well-to-do people occupy one and two-story houses with overhanging and tile roofs, while the poor class construct their mansions out of old boards, sugar cane stalks, barrel staves, pieces of matting, sun-dried bricks, and thatch them with palm leaves and dried strips of maguey. Their floor is always the ground. The highest temperature in the shade at Orizaba is 30 deg., the lowest 12 deg., but the average is mostly 21 deg., with always an east wind prevailing.

Orizaba is a delightful place for a stranger to stroll about in. We started out to see the town without guide or companion, and felt ourselves fully repaid by the many strange and delightful things we saw.

We went to the market, which is situated on an open square, and examined all the curious things. The birds especially attracted our attention, the many varieties, colors and shapes, and the extremely low prices, some selling for a medio (6 1/4 cents), others for a real. Young parrots were fifty cents, mocking-birds $1, and buglers, a bird shaped like the mocking-bird, but lighter in color and far superior in song, $2.50 and $3. All that restrains one from making a large investment is the fact that many cannot live in cages, as none know on what food they subsist, consequently they have to die. Little snow-white dogs, with bright black eyes and hair fine as silk, about three to five inches in length, sell for $2, while the famous Chihuahua dog, which weighs about half a pound when full grown, commands from $75 to $100, since tourists have ruined the prices.

Out by the unlucky Cathedral we saw the hearse of the town. It is the shape of a coffin, held aloft by springs above four wheels. It is draped with crape and plumes. Two black mules, stuck with plumes on every available spot, draw it, and the driver, dressed in black with high hat decorated with a plume, handles the reins, perched on a small seat about four feet above the rest of the hearse. The coffin is slid in at the back or end like the case in which coffins are often hermetically sealed.

Selecting a poor street, we started to make our way toward the mountains. On it we found a row of houses numbered in the following style: January 1, February 2, March 3, April 4, May 5, June 6, July 7, August 8. September 9, October 10, November 11, December 12. Still further down we saw one called "The place of Providence," each different door designated as "The place of Providence A, the place of Providence B," and so on throughout the alphabet. Next, we came to a laundry which did not remind us in the least of those at home. The river was the tub, a porous stone the washboard, and the little bushes and green bank the clothesline. In this manner all the washing of the town is done. We admired the washwomen for quite a while as they rubbed the clothes on the stone and then doused them up and down in the stream.

At last we concluded to jump across and go down on the other side, but we forgot we were women and that the dress of last fall was extremely narrow. We jumped from one washboard to another. We landed on it all right, but we did not stay long, but slipped back into the water, which was about three feet deep, much to our consternation. On our way home we stopped at the Tivoli, the bath-house and the main alameda, which is situated at the foot of an immense mountain, and is said to be one of the prettiest in the republic. The walks and drives are wide and nicely paved, a great variety of trees furnish the shade and musical fountains are plentiful. A music stand is in the center and is occupied nightly by a good band. The water-carriers were getting their supply from one of the large basins; they were also different from others we have seen. They have a long pole across their shoulders, and suspended from each end is a bucket containing the water, after the style of the milkmaids in the States. It seems strange that though every city has its water-carriers and that everyone in the same town carries exactly alike, yet in no two towns do they carry in the same manner.

I cannot forget to introduce you to the pleasant gentleman we met on the train. He is Mr. A. Baker, Her Britannic Majesty's Consul at Vera Cruz. He speaks fluently fifteen different languages, and when I asked him if he was not very proud of the fact, he replied: "Yes, until I met a waiter in a restaurant who could speak eighteen." He is a widower, and came here accompanied by his only son, while his three daughters are at school in Europe. The common expression made of him here is, that "he is good enough for an American." Now you can judge how agreeable he is. He has been knighted three different times, and was colonel in two different armies, yet he is still plain Mr. Baker. "Oh, I had ancestors," he said, jokingly, as we were discussing people's little vanities, "and they came over in the ship of the conquerors, also. My forefather was a cook. One day the bread was exhausted, and there was no way to procure more, so the cook made some pancakes, and waited in terror while they were taken in to his majesty. At last he got a summons to appear before him; trembling and expecting to be beheaded, the poor fellow sank at his sovereign's feet, when, instead of a sentence to be executed, he heard: 'Rise, Sir Baker.' Since then that has been the family name."

Accompanied by Mr. Baker, we started north to see a waterfall, and to take the train at the next station. We got in a car and went winding in between the high mountains from which the black marble is quarried until we reached a stretch of land, where we alighted and crossed the fields until we came to that wonderful stream. The water is quite cold and mineral, and as clear as crystal, one being able to see the bottom at the depth of twenty feet as though there was no water intervening. Down where the water was more shallow, were several horses fishing for the grass that grows in the bottom; they thrust in their heads until their eyes were in the water, and then pulled out a mouthful of grass; they made a beautiful picture. Baths are situated here, and trees grow around just plentiful enough to be pretty. Foot logs span the stream, and the cool, green, velvety plots invite a longer stay.

On one-foot log we discovered what appeared to be walking leaves, as the green leaves glided along, moved by an unseen power. Investigation proved them to be an army of ants, each one carrying a leaf on its back which looked like a little sail. On the edge of the bank, half in the water, half out, lay a branch of willow. These little things climbed, it, risking life and limb, and, cutting off a leaf, hoisted it on their backs and marched easily a quarter of a mile to their home. They had a path of road about five inches wide made along the grass all the distance. The street cleaners must be faithful, as it was as clean as could be, shaded on either side by the grass, without one blade in their way. They crossed the foot log and disappeared in a hole at the other end. We wondered what they were making inside with those many leaves. They were so interesting at their work that it was with reluctance we left them. Boarding our train, with much regret, we were soon lost to sight of the Valley of Orizaba and were once more on our way to a new and strange city.

CHAPTER XXII.
THE WONDERS OF PUEBLA.

IF the innocent-looking tourists believed all that is told them here they would conclude that every spot and town of interest had been built by the Virgin and the angels. One night many, many years ago, so the story runs, one good priest, who was known by the name of Motolinia, which means humble, mean, lowly, had a vision. A number of sweet angels – all of the feminine gender – draped with some soft, thin material, with long, silky black hair that fell to their feet in heavy folds, and sparkling black eyes, took the good father in their arms and bore him through the air to a spot not far distant from his little hut. After setting up a stone cross, which, at their petition, apparently descended from the skies, they helped him to build churches, houses, factories, and bull-rings (perhaps). It took seven days and the same number of nights to build the world, but the city of Puebla was built in a few moments. Probably the fatigue from work or the unusual company made the good man tired and drowsy, so he fell asleep, as sweetly as a babe, fanned by the wings of the heavenly beings around him. Waking from a most refreshing nap, in which he had dreamed of honey, golden crowns, feathered wings, and regiments of beautiful creatures, he found to his surprise that he was once again in his little bed, with no angels in sight. "They have gone out to complete the work, while I, lazy creature, slept," thought the good father, and going to the window he flung it open. He saw the green plain undisturbed.

At first, he was surprised and disappointed, and had he been a dyspeptic all would have ended there, and this story would not have to be told; but, like a good and faithful believer, he worked out a solution of the strange vision, which was that the angels had appeared to show they wanted the work done and how, but he must accomplish it himself. To prove beyond doubt their visit, the stone cross was left standing where their angelic hands had placed it. So, encouraged, and with great faith, he related his vision to the people, and with their aid began to build the city of Puebla around the stone cross, leaving more than a square vacant where it stood. This was three hundred and fifty-five years ago, on the 16th of April. The square is now used for the city market, and the stone cross, revered and respected, is standing in the courtyard of what was the convent of Santo Domingo, but now a church, where at the same place they will take from the altar and show you a coat which was once worn by a very holy monk, and for some good act the Virgin stamped her picture on the sleeve of it. It is very interesting to look at, even if one be so unfortunate as to possess but little faith; the most interesting thing in Puebla is its churches. Everyone has some wonderful tale attached.

Puebla was named in honor of its first visitors, Puebla de los Angles (Town of the Angels), but it is very seldom spoken of except as Puebla. The cornerstone of the first church was laid in 1531, and that of the first cathedral in 1536. Both of these buildings have disappeared, as they were originally, though it is proven that part of the former is the present Sagrario, covered with parasites and in almost utter ruins. The present cathedral was finished in 1649, and is one of the finest and most expensive in Mexico. One of its towers alone cost $100,000. The high altar, composed of Mexican marble and onyx, is one of the finest ever constructed. It is said to have cost $200,000. This altar, before the reform, was loaded with gold, silver, and jewels. The bishops' sepulcher is beneath. A beautiful carving in ivory of the Virgin, which was completed after three years' hard labor, and a wonderful curtain, which was a present from the King of Spain, as well as the dungeons beneath, are a few of the things worth seeing. It has eighteen bells. The largest weighs nine tons.

The Chapel of Conquistadora contains an image of the Senora de los Remedios, which was presented by Cortes to the Hascallan, Don Axotecatl Cocomitzin, for his good help and friendship during the time of the bitter war with the other natives. Upon the main altar lie the remains of the man who first introduced oxen into Mexico, and who for many years was the means of passage and communication between Mexico and Vera Cruz. His name was Sebastian de Aparicio. He was born in 1502, and died, after living a good and useful life, in 1600. At the Dominican Monastery they showed half the handkerchief on which the Virgin wept and wiped her eyes at the foot of the cross. The people claim that San Jose protects their town from lightning, so they built a church named in his honor, and have in it a strange image carved from what was a lightning-riven tree. Another beautiful church has a picture of a saint which has been heard to speak. Still another contains thorns from the crown of Christ. Nearly every two squares boasts a church, and every church has some wonderful history connected with it. The Church of San Francisco, which was founded by the good priest, Motolinia, the father of Puebla, was established a short time after the city, and is worth seeing, if from nothing else than an architectural point of view. The choir is the most wonderful thing in existence. It is flat and looks as though it would tumble down every moment; even the man who built it fled for fear it would fall, when taking out the supporting beams, and kill them all. The monks then decided to burn them down, and then if the choir fell no one would be hurt. Well, they burned and crumbled down, but the choir still remained firm, and does to this day, after at least two hundred and fifty years of constant use.

Puebla is fully seventy miles from the City of Mexico and is the capital of the state of the same name. It is one of the cleanest and prettiest towns in the republic, and has at the least 80,000 inhabitants. It is full of interesting historical events. Cortes located here; General de Zaragoza won a victory over the French here on the 5th of May 1862, and General Diaz, now President, won a more brilliant victory and gained greater fame for himself here in the war five years later, April 2, 1867. Both events are celebrated in fine style every year. Puebla is not situated on the main line of the Vera Cruz line, but connects with a narrow gauge which runs to Apizaco, twenty-nine miles distant. It takes from 4:40 to 6:10 to make the trip, but one forgets the length of time by looking at the beautiful valley which surrounds them. When we were out a short distance, by looking back over the way we traveled, we saw between two large hills, surrounded by trees, flowers and rocks, the Cascade del Molino de San Diego, showing just over the top of the falling waters a fine old stone mill inclosed with a variety of different green trees, all of which seem to be springing out of the waters whose fall faces us. Next, we pass the pretty little village of Santa Ana, interesting not only because it is named in honor of the old warrior, but for its people and the many odd things which they make so deftly and sell to passing tourists, Mexican drinks and ice cream, called agua a nieve (snow water), made simply by pouring sweetened and flavored milk over snow which is brought down from the Volcano Popocatapetl and the White Lady.

Between here and our destination we can see by the door of every little hut a large clay object shaped somewhat like an urn, taller even than the houses; they are, translated into English, "Keepers," and hold the water used by the people; they have no wells, as they carry their supply from a river many miles away, hemmed in on either side by a deep bluff. Although water is very scarce in the majority of places in Mexico, this is the only spot where one finds the keepers. Another town and we enter the city of Puebla, as it nestles down between a chain of mountains like a kitten in the sun. With a view from some high tower, it looks like a flower garden, dotted here and there with picturesque houses. On the west is Popocatapetl and Ixtaelhuatl, sending down an ever-cool and invigorating breeze, which plays with their snowy robes, and then descends into the green valley to salute the hot brows of mortals there with a kiss of health and happiness.

The coat of arms was given to Puebla by Charles V. of Spain, in July 1538. Of the inhabitants thirty thousand seven hundred are men and thirty-seven thousand eight hundred and thirty women, besides more than thirteen thousand people who work in public establishments, which number in all about one hundred. There are paper, cotton, flour and wax taper mills. The people are very religious, and fall on their knees when the bishop's carriage passes, even if it is unoccupied. They have plenty of policemen at night, though nearly everybody has retired by 10 o'clock, and not only are they on the streets, but on the housetops. We saw the little red lanterns blazing forth from almost every other house, and being of an inquisitive turn we made inquiries and learned the above facts. They look very odd, and on a dark night one can see nothing but the red light gleaming forth like a danger signal. The policemen are all well-armed, but it strikes an American that the lanterns are displayed so that their owners cannot accidentally get hurt. The city supports several free hospitals; the finest one was established a few years since, and is the best building of the kind in Mexico. Three days after the death of Luis Rharo, a bachelor of considerable wealth, they found in his Bible a will leaving one hundred thousand dollars to build this home and one hundred thousand dollars to be invested and used to maintain the same.

The three men named as executors – Clemente Lopez, A. P. Marin, and V. Gutiores – were all wealthy, but were to receive for managing and looking after the home, $15,000 apiece; the building alone was to cost $40,000, and after it was finished the contractor, E. Tatnans, would accept no pay and allowed the price to go back into the original sum. The building is marble, the floor marble tile, the decorations carved onyx, and this palatial mansion is to-day the home of hundreds of poverty-stricken and deserted mothers and babies. When Mexico feels charitably inclined she does it on a grand scale – no half-way business, like in many places in the States.

The houses here are generally two-story, with flat roofs, and fronts inlaid with highly glazed tile or else gaudily painted. All the windows facing the street have iron balconies, and all the courts are filled with flowers, birds and fountains. There were once seventy-two churches, nine monasteries and thirty nunneries, but the latter have been abolished, and, with, the exception of a half dozen, they are used as churches. One is a round house for the engines, another formed the theater for the bull ring. There are but two small Protestant churches, which are not well attended. Since the rebellion there have been established 1800 schools, with an attendance of only 36,000 children. The College of Medicine and Academy of Arts and Science are maintained at the expense of the town, free to all who care to go.

The famous pyramid of Cholula is but eight miles from the city. Street cars run out about four times a day and charge fifty cents, first-class, a trip. On the way we passed a large rock which has caused a sensation lately. It is about two hundred feet high and at the very least six hundred feet around the base. It looks very strange lying on an otherwise level green space for acres around. The stone is covered with parasitical orchids and ferns and has been known to the oldest natives by the name of Cascomate. No one ever thought much about it except to wonder how such an immense rock got into an otherwise rockless spot. Some advanced the opinion that it had been thrown there during one of the eruptions of Popocatapeti, when it merited the name of "the smoking mountain." A German who spends much of his time searching for the queer in Mexico thought one morning as he was taking a walk, about ten days ago, that he would climb to the top of this rock and take a view of the valley. The ascent was very difficult, but he persevered and on reaching the top was surprised to find a big opening yawning at his feet. The stench coming from it was very strong, so lighting his strongest cigar, he began to investigate. The opening was about fifty feet in circumference, and easily the same depth. At the bottom were lying several skeletons. He quickly returned to town and reported his discovery, but so far, no investigations have been made. One man who was talking over it said: "Please do not put it in your paper, because Mexico has a nasty name for foreigners now. That stone," he continued, "was used by duelists to hide their victim's body, and when the people perceived a stench they reported it to the police, who always investigated and had the body buried."

"If that is true, why is it that everybody considers the find new and startling, and no one has come forth to say he knows what use it was put to before this? If the police investigated and took out the bodies, why did they not have the hole filled up, and why are there so many skeletons in it at the present day?"

He did not try to answer these questions, but only begged our silence.

Cholula retains little of its old-time grandeur. At the commencement of the sixteenth century Cortes compared it to the largest cities of Spain, but with the growth of Puebla it has diminished, until the present day it is but a small village. Its streets are broad and unpaved, the houses one story with flat roofs, and there is little to attract one – although they have some few manufacturers – except the world-famous pyramid and some of the old churches. One of these churches was formerly a fortification built by Cortes. It is a fine, massive stone building of immense size. Perfect cannon of medium size answer for water-spouts on the roof. In the door of the main entrance there are 375 nails, no two of which are alike. When the building was being erected there were many skilled blacksmiths in the vicinity. Each was desirous of showing his skill, so with chisel and hammer they made these long nails and presented them to the conqueror, making the door one of the strange things of Mexico.

In another church near here, also erected at the command of Cortes, is a black velvet altar cloth, with saints embroidered in gold all over it. The workmanship is exquisite, and some of the likenesses perfect. There is also a black velvet vestment embroidered in the same manner, which is only for use in holy week. They were both a present from Charles V., of Spain. The Bishop of Mexico has been anxious to obtain possession of them, and has repeatedly offered $3500 for the two pieces, but they refuse to sell at that price. This church is known as the Royal Chapel. Its architecture is very pretty, yet extremely odd. Every way one counts across the chapel gives seven arches – lengthwise, crosswise, cornerwise, etc., the end is always the same – seven. In the center of this queer construction is a pure well, the waters of which are noted for their coolness, healing qualities and love charms. One strange fact about this church is that the morning following its dedication it fell to the ground completely demolished, but was immediately rebuilt. In this vicinity there are no less than twenty-nine churches, which can be counted, nestling within a very small space, from the pyramid, which is left for another chapter.

CHAPTER XXIII.
THE PYRAMID OF CHOLULA.

THE pyramid of Cholula is very disappointing to anyone who has seen illustrations of it in histories of Mexico; there it is represented as a mass of steps, growing narrower as they reach the top. At present it looks like many of the other queerly-shaped hills which one sees so frequently in Mexico. Closer inspection shows there were once four stories to it, but it is now badly demolished, and the trainway has cut through one side, damaging the effect. At present it is three thousand eight hundred and sixty feet around the base, although once it is said to have been one thousand four hundred and forty feet on, each side, or four times that around the entire base. Some say its height is no more than two hundred feet, while others affirm it is at the very least five hundred feet high; the ascent is made by a Spanish stairway of hewed stone fifteen feet wide, and there is a second stairway of two hundred steps leading from the main one to the church door.

The little church on top was first built by the Spaniards in the place of the temple called Quetzalcoate (the God of the air), built by the Astecs. The church was first in the shape of a cross, but alterations have been made of late years, destroying entirely the original design. It was dedicated to the Virgin of the Remedies, or Health – Senora de los Remedios, and she is said to have performed some wonderful miracles, at any rate her image is covered with tokens of her goodness. There is a desk in the church where they sell beads and measures of the Virgin's face, which are said to keep away the devil and bring good luck to the wearer. A little tinseled charm on the beads contains some part of the Virgin's garments, and when I, in a weak moment, asked the seller if he really meant it, I knew by his answer I had met George Washington, Jr. It was, "Senorita, I cannot lie."

At places where the hill is dug away can be seen the layers of mud-brick, which proves undisputedly that the pyramid was really built. It is thought to have served as a cemetery as well as a place of worship. The Indians have a tradition that when Cortes tarried at Cholula, a number of armed warriors plotted to fall suddenly upon the Spanish army and kill them all. Cortes may have had a suspicion, or a desire for more blood and more stolen wealth, for without the least warning, he attacked the citizens of Puebla and killed outright 6,000 besides terribly wounding thousands of others. When the road was being made from Puebla to Mexico they cut through the first story of the pyramid. In it was found a square chamber, destitute of outlet, supported by beams of cypress and built in an odd and remarkable manner. Curious varnished and painted vases, idols in basalt and skeletons were in it. The only conclusion offered was that it was either a tomb for burial or else the warriors who wanted revenge on the Spanish were by some means buried in this hiding-place. The pyramid is now covered with grass, trees and orchids.

Famous stone idols are found in this vicinity. In plowing the fields or digging holes they are turned up by scores, in all shapes and sizes; the tourist pays good prices for them, and the more sensational the story attached the higher the tariff; the guide at the hotel showed me a white arrow flint. He had bought it the day before at Cholula for a medio, and said he was going to daub it with chicken blood and sell it to the next party of tourists as a wonderful relic, which had been used on the sacrificial stone to kill thousands of people. He would tell them that the worshipers of the sun used to get a victim and the one who could send the arrow with this flint directly in the center of the victim's heart stood in favor with their god, the sun. At the depot, besides being bothered with at least twenty idol peddlers, a woman with a baby tried to make me buy it. She refused to sell to anyone in the party, but coaxed me to take it, telling all its good qualities. It was good, very amiable, sympathetic and very precious. Partly to get rid of her I asked, "How much?" "Dos reals" (twenty-five cents) was her astounding reply. "That is too cheap," I said; "I cannot take it unless the price is $100." Evidently, she did not understand jesting, for she kept on saying, " No, senorita, dos reals; muy benito." I successfully resisted its charms as well as her persuasions. At the last moment, when the car started, she ran after me, saying I could have the baby at $100, if I wouldn't take it at twenty-five cents; but the car soon left her in the distance, and we had a good laugh at the poor woman's reasoning powers and lack of business qualities.

The tramway ends at Atlixco, a lovely little village midway between Cholula and Puebla. One of the most beautiful things along the way is the famous tree at the foot of St. Michael's Mountain. It is called Ahuehuete. It is many centuries old and a very curious shape. Its trunk is hollow, with a hole big enough for a horseman to enter at one side. Thirteen men on horseback can find plenty of room in its big body. The orchards at this village are valued at $2,800,000.

There are twenty-four hotels in Puebla, and some are first-class in every respect. They serve coffee from 6 to 9, breakfast 1 to 3, and dinner 6 to 9. The penitentiary looks like a Spanish fortress. It is very old, picturesque, and covered with orchids, but the state authorities decided they needed a new one, and have built a handsome one of stone and brick, which is said to resemble one in Pennsylvania, whether East or West I know not, but from a distance it looks somewhat like the Western, although all similarity faded on closer inspection. There are several parks, and very pretty ones, too, in Puebla. In the main one they have music nightly. At the east end of the town they have sulphur baths, which are considered very healthy.

The most unique bull-fights of the whole republic are held here. One Sunday they fought all afternoon in the regular style, but when evening came, they turned on the electric lights, set a table in the center of the ring, put on it tin dishes, and all the fighters sat down as though to eat, one of them attired in a long, white dress. As soon as they were seated comfortably the gate was flung open, and the toro rushed in. At the same moment two banderillas containing fire-rockets were stuck into him, and as they exploded the maddened bull made a rush for the table. The occupants jerked up the tinware, and with it began to fight off the bull. Then they jerked the table apart, and fought it with the pieces. When the men and beast were pretty tired, the bull was allowed to attack the one in white, the so-called bride, and the swordsman, who of course represented her husband, defended her, and killed the bull with one thrust of the sword. It was simply magnificent, and so exciting that everybody was standing on their feet yelling lustily at every new move. The fight was called "The Interrupted Bridal Party."

The next Sunday they fought the bulls on burros instead of horses. The men had their bodies protected by plates of tin, and when the toro charged they jumped off the borro and ran behind screens, while the poor little animal had to run for his life, and that was the funniest part of the programme. The following Sunday all the fighters stuffed themselves. They looked as if they had feather beds around their bodies. Then they dressed up in fantastic garb. No horses were allowed in the ring. When the time came the men lay flat on their backs, and and as the door was opened and the bull came tearing in, they wiggled their legs in the air to attract its attention.

One peculiar feature of bull-fighting is that the bull will never attack a man's legs, but always strike for his body. The toro would rush for the prostrate form, and the American auditors would hold their breath, and think that the fighter's end had come, but just then the bull would gore him in the stuffed part, and the man would turn a complete somersault, alighting always on his feet, safe and sound. The bull would turn those men into all sorts of shapes without either hurting them or himself.

Puebla is considered the richest State in Mexico, and in it one can select any climate he desires. Puebla City is never cold, is never warm; it has the most delicious climate in the world, just the degree that must please the most fastidious. In the State are wonderful stone quarries. Every color of clay is used to make dishes, vases, and brick, and abundance of chalk for making lime. In the rivers and small streams several kinds of sand are secured, which is used for many purposes, and a few miles away are large veins of iron and other minerals; there are mountains of different varieties of marble and onyx, from the transparent to the heaviest known; extensive fields of coal, quicksilver, lead, with wonderful mines of gold and silver everywhere; there is one strange mountain called Nahuatt (star) covered with rock crystal, the fragments resembling brilliant diamonds, and at another craggy place beautiful emeralds are found. In many places are hot springs.

The woods are fortunes in themselves. Besides all the Mexican varieties are cedar, ebony, mahogany, pine, oak, bamboo, liquid amber, India rubber, and above all the writing-tree, the wood of which has been pronounced the finest by five countries. Its colored veins are on a yellowish ground, and it forms thousands of strange figures, monograms, words and profiles. Then there are the silk cotton tree, the logwood and thousands of others. Some of them produce rich essences, others dyes, which never fade. A cactus also grows here from which wine is made which they say far excels that of Spain or Italy. In the cold and warm districts are raised cotton, tobacco, vanilla, coffee, rice, sugar-cane, tea, wheat, anise-seed, barley, pepper, Chili beans, corn, peas, and all the fruits of the hot and cold zones. There are salt mines and land where cattle, horses, mules, burros, sheep, goats and pigs are raised on an extensive scale. The flowers are so varied and abundant that a gentleman who has been exploring the paradise says their products would supply all the drug stores of the world with perfume. These are a few of the charms of the State of Puebla.

There is quite an interesting story connected with the emerald district. The Indians found one and placed it on the altar of the church to serve as a consecration stone. It was three-quarters of a Spanish yard, or a little over one-half English yard, in length. Maximilian, during his short reign, went to Puebla to examine it, and offered $1,000,000 for it the moment the jewel expert with him pronounced it extremely fine. The Indians refused, and asked $3,000,000. Afterward an armed force went to kill the tribe and carry off the gem, but were themselves whipped. The Indians then decided to bury it for safe keeping, when a wily Jesuit promised eternal salvation to the living, the dead, and the unborn, if they would give it him in the name of the Holy Virgin, who, he said, had asked for it. The poor innocent and faithful wretches gave their immense fortune away for a promise that was worse than nothing, and the treacherous purchaser cut it into small portions and sent it across the sea to be sold, he, reaping the benefit. The god Quetzalcotl, which once graced the top of the pyramid at Cholula, was sold to an American a few years since for $36,000.

A few miles out from the city, situated in the midst of a barren plain, stands the magnificent old castle of Perote, which is celebrated in Mexican history as the last home of many of her dark-eyed senoras, who have either pined to death in its dreary dungeons or been murdered during revolutions. It was once the national prison of the republic, and was considered one of the strongest buildings in the world. Even now it is stronger and more formidable than most fortresses. There is much more of interest, historical and otherwise, to be seen in and around Puebla, and one could spend mouths of sight-seeing every day, and still have something worth looking at. If a gentleman or lady resident of Puebla is asked where their home is they will quickly answer, "I live in Puebla, but am not a Pueblaen." The latter word translated into Spanish means false and treacherous, hence the carefulness of the people always to add it.

I cannot end this until I give you a sample of the meanness of the Mexicans, other than Indians. The real Mexican – a mixture of several nationalities – has a great greed for cold cash, and thinks the Americano, Yankee, or gringo, was sent here to be robbed. They do not draw the line on Americans, but also rob the poor Indian of everything. When I asked for my hotel bill, which was $4 a day, the clerk handed me a bill with $1.25 extra. "What is the extra for?" I inquired. "Charming senorita," he answered, "you called for eggs two or three times." "Yes," I replied, "when you set down goat's meat for mutton, and gave me strong beef I had seen killed by the matadore in the bull-ring the day before." "Well," he continued, "eggs are expensive, and it was a trouble to cook them." "My dear senor, I have no intention of paying your salary, and your pocket is just minus an expected $1.25. Here is the other." That settled it.

While looking at some marble objects in a store a poor Indian came in with twelve blocks of marble twelve by twelve, on his back; the poor fellow had hewn them smooth and then traveled undoubtedly two days or more on foot over hills and through valleys, the ground at night his bed and the wild fruits or a few beans brought from home his food. He was ragged and tired, and dirty, but he had a good, honest look on his face. He asked the shopkeeper to buy the marble. After a little inspection the merchant purchased, and for it all, which was weeks of labor to the poor peon, and meant at least $300 for himself, he gave fifty cents. Nor was that the worst of it; the two quarters were counterfeit and the Indian told him so, but he said no. I stepped to the door and watched the peon go to a grocer's store across the street. They refused to take the money and he came back and told the marble dealer. Upon his refusing to give good money the Indian turned to me for help, whereupon the keeper laughed and said: "She is a Yankee and can't understand you."

Well, I had not been in Mexico long, and was entirely ignorant of the language, but my American love for justice was aroused, and in broken English and bad Spanish I managed to tell him I knew the money was bad, and that the merchant was like the money – that by even giving good money he was cheating the poor peon of his goods, He was surprised, that is if a Mexican can be surprised, and he gave out some little change, which I examined, and not being sure whether it was good or bad, put it into my own purse, giving the man a quarter instead. He thanked me warmly, tied the money up in the corner of a rag he had tied around his waist, and then went out and tried the other quarter. This also failed to pass, and he returned to the now furious storeman, who threatened to call the police if he did not go away. "If you do, I will tell them that you are passing counterfeit money," I said, whereupon he gave the peon another piece, and the poor fellow departed happy. While the storekeeper said some nasty things in Spanish about "Gringos," it is needless to add I did not buy, nor had he the least desire to sell to me.

CHAPTER XXIV.
A FEW NOTES ABOUT MEXICAN PRESIDENTS.

VERY few people outside of the Republic of Mexico have the least conception of how government affairs are run there. The inhabitants of Mexico – at least it is so estimated – number 10,000,000 souls, 8,000,000 being Indians, uneducated and very poor. This large majority has no voice in any matter whatever, so the government is conducted by the smaller, but so-called better class. My residence in Mexico of five months did not give me ample time to see all these things personally, but I have the very best authority for all statements. Men whom I know to be honorable have given me a true statement of facts which have heretofore never reached the public prints. That such things missed the public press will rather astonish Americans who are used to a free press; but the Mexican papers never publish one word against the government or officials, and the people who are at their mercy dare not breathe one word against them, as those in position are more able than the most tyrannical czar to make their life miserable. When this is finished the worst is yet untold by half, so the reader can form some idea about the Government of Mexico.

President Diaz, according to all versions, was a brave and untiring soldier, who fought valiantly for his beautiful country. He was born of humble parents, his father being a horse dealer, or something of that sort; but he was ambitious, and gaining an education entered the field as an attorney-at-law. Although he mastered his profession, all his fame was gained on the battle-field. Perfirio Diaz is undoubtedly a fine-looking man, being what is called a half-breed, a mixture of Indian and Spaniard. He is tall and finely built, with soldierly bearing. His manners are polished, with the pleasing Spanish style, compelling one to think – while in his presence – that he could commit no wrong; the brilliancy of his eyes and hair is intensified by the carmine of cheek and whiteness of brow, which, gossip says, are put there by the hand of art. Diaz has been married twice – first to an Indian woman, if I remember rightly, who left him with one child, and next to a daughter of the present Secretary of the Interior, Manuel Romero Rubio. She is handsome, of the Spanish type, a good many years younger than the president, and finely educated, speaking Spanish, French and English fluently. Mrs. Diaz has no children, but is step-mother to two – a daughter and a son of the president. The president, so far as rumor goes, follows not in the footsteps of his countrymen, has no more loves than one, and is really devoted to Mrs. Diaz.

There are two political parties, a sort of a Liberal and Conservative concern, but if you ask almost any man not in an official position he will hesitate and then explain that there are really two parties; that he has almost forgotten their names, but he has never voted, no use, etc. Juarez, who crushed Maximilian, while a good president in some respects, planted the seeds of dishonesty when he claimed the churches and pocketed the spoils therefrom. Every president since then has done what he could to excel Juarez in this line. When Diaz first took the presidency, he had the confidence and respect of the people for his former conduct. They expected great things of him, but praise in a short time was given less and less freely, and the people again realized that their savior had not yet been found. When his term drew near a close, his first bite made him long for more, and he made a contract with Manuel Gonzales to give him the presidency if he would return it at the end of his time, as the laws of Mexico do not permit a president to be his own successor, but after the expiration of another term (four years) he can again fill the position.

The constitution of Mexico is said to excel, in the way of freedom and liberty to its subjects, that of the United States; but it is only on paper. It is a republic only in name, being in reality the worst monarchy in existence. Its subjects know nothing of the delights of a presidential campaign; they are men of a voting age, but they have never indulged in this manly pursuit, which even our women are hankering after. No two candidates are nominated for the position, but the organized ring allows one of its members – whoever has the most power – to say who shall be president; they can vote, though they are not known to do so; they think it saves trouble, time, and expense to say at first, "this is the president," and not go to the trouble of having a whole nation come forward and cast the votes, and keep the people in drunken suspense for forty-eight hours, while the managers miscount the ballots, and then issue bulletins stating that they have put in their man; then the self-appointed president names all the governors, and divides with them the naming of the senators; this is the ballot in Mexico.

Senor Manuel Gonzales readily accepted Diaz's proposition and stepped into the presidency. He had also been a loyal soldier, and was as handsome as Diaz, though some years his senior. Gonzales is a brave man, powerfully built, but was so unfortunate as to lose his right arm in battle. He has, however, learned to write with his left in a large, scrawling style. He has a legal wife, from whom, however, he is separated. While he was filling the presidential chair, she made a trip through the United States, and gained some notoriety by being put out of the Palmer House because she did not pay bills contracted there on the strength of being the wife of the President of Mexico. On her return to the land of the Aztecs, she found that the law could not touch the Czar Gonzales, who was living like a king, nor could she get a divorce, as Mexico does not sanction such luxuries. She started a sewing establishment, but it is said that she is living in abject poverty, and, like all Mexican women, with the door to the way of gaining an honest livelihood barred against her because of her sex.

Their family consists of two sons, both captains in the army – Manuel, twenty-seven years old, and Fernando, twenty-five – fine-looking and well educated. The latter is said to be quite good to his mother. It is reported that Manuel Gonzales and Miss Diaz, the only daughter of the president, are to be married shortly.

Gonzales while in power issued several million dollars' worth of nickel money, which the people refused to accept. One day, as he was being driven from the palace in an open carriage, he was surrounded by a mob who threw bags of the coin on him, while others cried out for his life. The driver – who, by the way, was at that time the only negro in the City of Mexico – fiercely fought those who had stopped his team and resisted by main force their efforts to unseat him. He wanted to drive the fine-blooded horses right over the angry, howling mob, but Gonzales calmly told him to desist, and then, revolver in hand, descended from the carriage, asked the people what they wanted, swore roundly at them and commanded them to disperse.

The effect was astonishing. Without one outburst, as though quelled by an immense army, that maddened mob moved away and Gonzales re-entered his carriage triumphantly, and was driven home unmolested and uninjured. The money, however, was sold for almost nothing, and some Europeans were smart enough to buy. In a short time, the government bought it all back, paying cent for cent, and I know personally one man who made $100,000 in one day on his lot. Ill truth, it was the foundation of more than twenty fortunes in Mexico at the present time. Eight months before Gonzales retired he tried to force the people to accept the English debt law. They refused, and filled the halls of Congress, in which they had congregated, with cries and groans. They would not cease at the presidential command, and Gonzales ordered the soldiers to fire on them several times. It was impossible that in such a narrow space all should escape death, yet no true report was ever made of the affair.

When Gonzales went into office $900,000 could be counted in the treasury. On the last day of his term his annual income exceeded $200,000 and his salary, which was $30,000 yearly. On the morning of his last day he sent to the treasurer to know how much money yet remained in the treasury. "One hundred thousand dollars," was the reply. Gonzales requested that it be sent to him, and when the treasurer meekly hinted that it might be good for his neck to know to whom to charge it, Gonzales replied that if he did not know that much he had better send in his resignation. The money was in the president's hand in a very short time after this. Next, he bought a $2 ticket from the state national lottery and with it sent a little line to the managers. "See that this draws the prize to-day." The first prize was $100,000. Strange to relate his ticket drew the fortunate number, and Gonzales closed his eyes that night with a murmur like Monte-Cristo as he gazed upon the sea, "The world is mine!" That evening the people were so glad that they gathered in an impassable mob around the palace and cathedral, and tried to enter the latter, that they might proclaim their feelings by ringing forth from the numerous bells which hang in the mammoth towers, one happy peal; but an army was soon on the spot and prevented any demonstration. Investigation showed $25,000,000 missing and the government employes unpaid.

Experts figure out that Gonzales raked in $25,220,000 in his four years of official life, and he didn't have to go to Canada, either. Gonzales immediately went to Guanajuato as governor, where he was received with open arms, and when the people, who found the bank broke just as they expected to take it, began to whisper that they would like a little investigation, Gonzales swore he would spend every cent they were clamoring after in raising an army to overthrow the Diaz Government. On hearing this Diaz slunk off like a half-drowned cat and made a law, which went into effect June 22, 1886, taking a percentage off every government employe to help pay up the Gonzales deficiency.

Gonzales is modest; he don't want the presidency any more. He wisely invested his hard-earned cash in an estate. His palaces and haciendas are something wonderful for size, beauty, and furnishment. Of course, give a man a bad name and everything mean is laid at his door; but it is credited to him that he took a fancy to a very rich hacienda, and he told the owner he would give him $200,000. The haciendado said it had belonged to his family since the time of Cortes, and he had not the least desire to sell, besides it was at the very least worth $2,000,000. Immediately all sorts of evil fell upon the unhappy owner. His horses were shot, his cattle, water, and even family poisoned. At last, when hope was crushed, Gonzales accidentally reappeared, and told the heart-broken man that he would give him $10,000 for this place. The hacienda was immediately his, but the former owner is still looking for his money. The strange part is that Gonzales has not suffered the afflictions visited upon the former owner.

President Diaz has two years from next December to serve, that is, providing a revolution does not cut his term short. The people will not say much about his going out, as one just as bad will replace him. They always know one year in advance who the president is to be, and even at the present date it lies between Diaz's father-in-law, Romerio Rubio, or Mier Teran, Governor of Oaxaca, both of whom belong to the ring. Diaz fears a revolution, and is afraid of losing his life. It is said he hastened his removal to Chapultepec because they threatened to blow up his house on Calle de Cadena, No. 8, with dynamite. Last January a party of Revolutionists laid plans to overthrow the Diaz Government, but one fellow got into a controversy with a Diaz party while riding on the Pasio, and so they came to blows, the news got abroad and armies paraded through the streets of Mexico until the poor little body of "righters" were overawed by the demonstration. Gonzales is sixty-five years old. He gets along nicely as Governor of Guanajuato, having no duties and being looked up to as a king by the people. When he comes to Mexico for a few days they prepare expensive receptions for his return. They are his humble subjects, and he is satisfied to be king of that state.

CHAPTER XXV.
MEXICAN SOLDIERS AND THE RURALES.

EL MEXICANO thinks it would be one of the pleasantest, as well as one of the easiest, things in the world to whip the "Gringoes," while the latter, with their heads a little swelled, perhaps, imagine otherwise, and scoff at the idea of the "Greasers" winning even one battle in the event of war. Be that as it may, solid, unvarnished facts will prove to the most headstrong that the advantage is mostly on the other side.

The standing army in Mexico is said to number forty thousand men, but is believed to be more. Every little village of a few hundred people has its army, and every day that army is being increased; the officers range from those who have gained experience and fame on the battlefield to the young ones reared and trained in military colleges; they are mostly all of what is considered the highest class of people in Mexico.

The rank and file are mostly half-breeds or Indians, who are not by any means volunteers. They are nearly all convicts. When a man is convicted of some misdemeanor he is enlisted in the regular army, separated from his home, and to serve the rest of his natural life. This life is not a bed of roses – there is no bed at all, and out of a medio (6 1/4 cents) a day, he has to furnish his food and comforts. The dress uniform is made of coarse woolen goods, with yellow stripes on the sleeves; and the undress uniform, which is worn constantly except on review days, is but white muslin, pants, waist and cap.

Some of the Indians are stolen and put in the army, and they immediately resign themselves to their fate, for there is no more escape for them than there is from death.

The wives of these poor fellows are very faithful, and very often follow the regiment from one place to another; they live on what nature grows for them and what they can beg or steal; the men are called in Spanish "soldados," and the women, because they cling to their husbands, "soldadas." It looks very pitiful to see a poor Indian woman with a babe tied to her back and one clinging to her skirts, dusty, hungry and footsore, traveling for miles through the hot sun with the regiments.

These soldados are wonderfully hardy; they can travel for a week through the hot sun, with nothing to drink and but a spoonful of boiled beans and one tortillia – a small flat cake – for two days' rations, sleep on the ground at night, and be as fresh for service as a well-kept mule. Fight! well those who imagine it such an easy thing to whip them should stand off and witness some of their feats first; they love their country, and consider life well lost in defense of it; they are ignorant, it is true, but seem the more courageous for it. When told to fight, they go at it with as much vigor as a bull dog after a cat; they don't know why they are fighting, or for what, but it is their rule and custom to obey, not to reason why. If you would stop one soldier in the midst of his fighting and ask: "Why are you fighting?" he would answer in the characteristic words of his people, "Quieu sabe?"

If a man is silly enough to try to escape from this bondage he is immediately shot, or if he disobeys orders they have time but to punish him with death. A short time before leaving Mexico some guards at the prison tried to desert, and immediately every regiment was notified to be on the lookout, and others were sent out to recapture them, and as soon as found they were shot. The soldiers have an herb named marijuana, which they roll into small cigaros and smoke. It produces intoxication which lasts for five days, and for that period they are in paradise. It has no ill after-effects, yet the use is forbidden by law. It is commonly used among prisoners. One cigaro is made, and the prisoners all sitting in a ring partake of it. The smoker takes a draw and blows the smoke into the mouth of the nearest man, he likewise gives it to another, and so on around the circle. One cigaro will intoxicate the whole lot for the length of five days.

The Mexican officers are unpleasantly sarcastic, or rather they have a custom that is the extreme of irony. It is known as la ley fuga (the law of escape). They will tell you they are going to take a prisoner, or soldier, as the case may be, out to the suburbs to give him a chance to escape. It sounds very pleasant to the stranger. They will, for example, politely ask the railway conductor to stop the train in some quiet place, as they want to let a prisoner escape. The American conductor finds his heart warming within him for these generous officers, and quickly and gladly obeys. The train is stopped, they all get off, and the officers form in a single line, with guns raised to the shoulders. The prisoner is placed before them and told to *Vamos*. He gives one glance into their unchanging faces, the surrounding land, and then starts. That moment he falls to the earth riddled with a dozen bullets, and the executioners re-enter the train and are speeding fast away, almost before the echo of this fatal volley died away. They cannot waste time putting his body beneath the ground, but before long, some Indians, traveling that way, find it. He is one of them, and their turn may be next, so they lay him in a hastily-dug hole, erect a wooden cross at the end, murmur a prayer, and leave him to return to that from which he sprung. This is the merciful "law of escape" practiced daily in Mexico.

Once every year to commemorate the victory over the French on the 5th of May 1862, the president reviews all the troops. They flock to the city from mountain, valley, town, and city, clad in holiday attire. Then only one realizes their strength, as they march before the palace where the president is seated on the balcony. The finest looking men in the whole 40,000 are the rurales. They number 6000 and are larger men than Mexicans usually are.

These rurales are a band of outlaws who came forward with their chief and aided Diaz during the war. When it was over Diaz recognized their power, and was so afraid of them that he offered them a place in the army, with their chief as general, and they are to-day not only the best paid, but – speaking of their fighting ability – the best men in Mexico. In the first place they are large and powerful and known over the entire country, mountain, town, and valley, as thoroughly as we know our A, B, C. They fear nothing on earth, or out of it, and will fight on the least provocation. They would rather fight than eat, and have a great aversion to exhibiting themselves, as they demonstrated on the 5th of May last, when only 800 could be persuaded to participate.

They have their own bands and a number of buglers. Every man owns his horse, which must in color match that of the rest of the regiment. Their uniform is yellow buckskin, elaborately embroidered with silver and gold, upon the pants and on the back, front and sleeves of the short cutaway jacket. Their wide sombrero is the same color, finished with the same embroidery and a silver cord and tassel. Their saddles also match their suits in color and silver finish. How they ride! It is simply perfection. The horse and rider seem to be one.

I don't think they could carry any more weapons if they tried. Each man has a good carbine, a sword, two revolvers, the same number of daggers and two lassos, and they fight with any or all of these weapons. They fight very cleverly with the lasso. If they wish to take a prisoner – a very unusual proceeding on their part – they, with the rope, can either lasso man and horse together or two or more men. The other lasso is of wire, which not only catches the fugitive, but knocks him senseless or cuts his head off, as the case may be.

These rurales guide tourists through the interior and also attend all public places to keep order; they receive one dollar a day, which is enormous compared with the other soldiers' pay of six and one-quarter cents. They have their horses in perfect control, and can make them execute all kinds of movements in a body, while the tricks performed by individual horses are numberless.

The Mexicans have a good deal of suppressed wrath bothering them at the present day; they know that Diaz is a tyrannical czar, and want to overthrow him. It may be readily believed that Diaz knows they are bound to get rid of this superfluous feeling, and he would much rather have them vent its strength on the Americans than on himself; thus, he stands on the war question. He is a good general, and has many good, tough old soldiers, the best of whom is ex-President Gonzales, to aid him, besides the convict soldiers and the rurales.

CHAPTER XXVI.
THE PRESS OF MEXICO.

THE press of Mexico is like any of the other subjects of that monarchy, yet it is a growing surprise to the American used to free movement, speech and print who visits Mexico with the attained idea that it is a republic. Even our newspapers have been wont to clip from the little sheets which issue from that country, believing them untrammeled, and quoting them as the best authority, when, in truth, they are but tools of the organized ring, are only capable of deceiving the outsider.

In the City of Mexico there are about twenty-five newspapers published, and throughout the empire some few, which are perused by the smallest possible number of people. The Mexicans understand thoroughly how the papers are run, and they consequently have not the slightest respect in the world for them. One can travel for miles, or by the day, and never see a man with a newspaper. They possess such a disgust for newspapers that they will not even use one of them as a subterfuge to hide behind in a street car when some woman with a dozen bundles, three children and two baskets is looking for a seat.

The best paper in Mexico is *El Monitor Republicano* (the Republican Monitor), which claims to have, in the city, suburbs, and United States, a circulation of five thousand. It is printed entirely in Spanish. The *Mexican Financier* is a weekly paper – filled with advertisements from the States – which is published in English and Spanish, and is bought only by those who want to learn the Spanish language, yet it is the best English paper in Mexico. Another English paper is published by an American, Howell Hunt, in Zacatecas, but it, like the rest, is of little or no account. One of the newsiest, if not the newsiest, is *El Tiempo* (the Times), which is squelched about every fortnight, as it is anti-governmental.

Very few have telegraphic communication with the outside world, and none whatever with their own country. They mostly clip and translate items from their exchanges, heading them "Special telegrams," etc., when in reality they are from eight to ten days old. *El Monitor Republicano* steals from its exchanges first and the other papers copy from it. Not a single paper has a reporter. Two men are considered plenty to clip and translate for a daily, and it is not unusual for them to borrow type to set the paper. All the type-setting is done in the daytime and a morning paper is ready for sale – if anybody wanted it – the afternoon before. While our morning newspapers allow their brains to rest at 5 A. M., the Mexican brethren cease labor the day before at 4 P. M. Things happening on the streets, which would make a "display head" with us, are never even mentioned by them. One day I saw woman fall dead two squares away from a newspaper office, and after a long time read in the same paper: "One of our respected contemporaries is authority for the story than an unknown Indian woman dropped dead on the street about two weeks ago." It needed no label "castanado" (chestnut). For a time, the papers imagined they had an item.

There was an old Frenchman who made some sort of taffy and with it used to perambulate the streets crying, "Piruli." The English paper came out quoting a notice of this old fellow. In a few days they quoted another to the effect that the old fellow had died of smallpox. Then, after using space for one entire week, changing every other day the cause of the old man's death and substituting some new disease, the learned editor stated that according to all reports the old fellow was not dead at all, but had charmed some rich Mexican widow with his musical voice – or taffy – and was enjoying a honeymoon on her bank account. We even did not get peace with that, but in a few days, they declared the report false and gave a new version. When we left there, five months later, they were still contradicting themselves about the old taffy-peddler.

Quite as bad was their treatment of a small forest fire located about twenty miles from the city. I was at the village at the time, and was quite amused, when the fires were extinguished after eight hours' burning, to read for two weeks after contradictory stories on it. It was still raging with renewed energy – hundreds of lives had been lost, etc., until one morning the English paper said: "According to a letter received at this office yesterday, the forest fire only lasted a few hours, and our contemporaries, from whom we have been quoting, have made a big mistake. No lives were lost."

When a new member was added to the royal family of Spain the notice was clipped from a foreign paper, in which it stated clearly that the Queen Regent Christina had given birth to a boy baby. Yet it was headed: "Is it a Boy?" When it grew a little colder than usual in an interior town, they headed the item: "A Mexican Town in Danger." When Roswell P. Flower, of New York, returned from his trip to Mexico he was interviewed by some reporter, and while he said nothing in Mexico's favor he said nothing against it, so they headed the clipping: "He Loves Mexico." Moralizing is quite customary, at least with the English paper. After quoting an item from *La Patria* about a married pair quarreling so fiercely that the mother-in-law took bilious fever and died, it gave a sermon entitled: "Let not your angry passions rise." On another occasion, speaking of the criminal list being unusually large for the last month, it broke out with: "Oh, pulque, pulque, what evils are committed under thine influence! And yet, verily, thou art a most excellent aid to digestion."

All the papers which I know of are subsidized by the government, and, until within several months ago, they were paid to abstain from attacks on the government. This subsidy has stopped, through want of funds, but the papers say nothing against the government, as they care too much for their easy lives; so, they circulate among foreigners misrepresenting all Mexican affairs, and putting everything in a fair but utterly false light. The Mexicans have nothing but contempt for the papers, and the newspaper men have no standing whatever, not even level with the government officials, whose tools they are. If a newspaper even hints that government affairs could be bettered, the editors are thrown into prison, too filthy for brutes, until they die or swear never to repeat the offense. The papers containing the so-called libelous items are all hunted up by the police and destroyed, and the office and type are destroyed. These arrests are not unusual; indeed, they are of frequent occurrence. While in Mexico I knew of at least one man being sent to jail every two weeks; they are taken by force, in the most peculiar manner for a country which lays claim to having laws, not to speak of being a republic. Just for an imaginary offense in their writings, they are remanded to prison, and are kept in dark and dirty cells, shut off from connection with the world without trial, without even enough to eat.

A satirical paper named *Ahuizote* was denounced by some offended government officials and the editor was thrown into jail. Then Daniel Cabrera started another Mexican Puck and called it *Hijo del Ahuizote* (the son of Ahuizote). It was quite clever and got out a caricature entitled: "The Cemetery of the Press," showing in the background the graves of the different papers, and in the front a large cross engraved, "The independent Press. R. I. P.," while hanging to each side was a red-eyed owl with a spade. On top of the tomb was a lighted fuse marked "Liberty." Underneath it read, "The sad cemetery of the Press of Mexico, filled by liberty leaders, Juarez, Lerdo, Diaz and Gonzales." The police were sent out to gather up and destroy every copy of this paper.

Editor Cabrera was put in Belem, where he remained in the most pitiable condition until death promised release; through the influence of friends they took him home to die, guarding his house with a regiment until he should be fit to be carried back to jail or until they should see his body consigned to the grave. To say libelous things is as dangerous as to write them. One fellow who ran a liquor shop let his tongue wag too much for wisdom, and one night, a member of the police secret service went in, and as the proprietor turned to get the drink the policeman had called for, he was shot in the back and again in the body after he had fallen. The notice of the affair ended by saying: "It is not known whether the policeman had orders to do the shooting." *La Cronicade Tribunales* (the *Court Chronicle*) editor was denounced and imprisoned for simply speaking about the rulings of one of the judges.

As all know by the Editor Cutting case, even a foreigner does not write about Mexico's doings as they really are. I had some regard for my health, and a Mexican jail is the least desirable abode on the face of the earth, so some care was exercised in the selection of topics while we were inside their gates. Quite innocently one day I wrote a short notice about some editors, who received no pay from the government, being put in jail. The article was copied from one paper to another, and finally reached Mexico. The subsidized sheets threatened to denounce me and said in Spanish, "One button was enough;" meaning by one article the officials could see what my others were like, but by means of a little bravado I convinced them that I had the upper hand, and they left me unhurt. They have a law, known as "Article 33," which defines the fate of "pernicious" foreigners who speak or write too freely of the land and its inhabitants. Once or twice they have been kind enough to take the offending foreigner and march him, with a regiment of soldiers at his heels, across the boundary line.

Professor Francis Wayland, of Brown University, together with the American Consul, Porch, and Dr. Parsons, visited the prison Belem to ascertain the conditions of the editors imprisoned there. They were not granted any of the customary privileges, but one little paid sheet was afraid some truth would reach the public's eye, as Professor Wayland was soon to return to the States. In referring to the visit, this paper said: "It is to be noted that these men wanted to enter the very gallery where the newspaper men were confined, and that they took 'note in a memorandum book of all answers.'" To save trouble, Dr. Parsons, who resides in Mexico, said they merely exchanged the usual greeting with the prisoners. Some of the editors confined thought, that as they belonged to a press club, that they could appeal to the Associated Press of the United States for aid. Of course, such an appeal would be useless; the papers now published there take pride in copying and crediting them to other papers. No dependence can be put in any of them for a true statement of affairs. The *Two Republics* was started and run by a Texan, Major Clarke. He lived in Mexico with his family and regularly every evening used to take a walk down the paseo with his two daughters, who always walked a couple of yards in advance. This was repeated every day until the Mexicans used to say, "There is Clarke and his Two Republics."

CHAPTER XXVII.
THE GHASTLY TALE OF DON JUAN MANUEL.

WHEN able to translate Spanish, there is nothing that will amuse a tourist more in the City of Mexico than rending the street and store signs and names of the different squares. Streets are not named there as here. Every square is called a street and has a separate name; the same with all the stores and public buildings. No difference how small, they have some long, fantastic name painted above the doorway. We used to get lunch at a restaurant called "The Coffee House of the Little Hell," and our landlady always bought her groceries at "The Tail of the Devil."

"Sara's Shoe," the "Paris Boot," and the "Boot of Gold," were all shoe stores of the very best order, where they will make lovely satin boots, embroidered in gold or silver bangles for $8 a pair, or of the finest leather for $3 to $5. They never have numbers to their shoes, and if none will fit, they make to order without extra charges. There is not a low-heeled, flat shoe in Mexico; they cannot be sold. One pair of American make, in a window on a prominent street, attracted a great deal of attention and ridicule. The Mexican women have lovely feet, and their shoes are very fancy – extremely high cut, French or opera heels and pointed toes. The shoemakers have a book in which they take orders for shoes. First, they set the foot down on a clean page and mark out the exact size; then they write on it the measure and the thickness, and when the shoe arrives it is of perfect fit. Let it be added, as encouragement to La-Americana, that although the dark-eyed Senorita's foot is exquisite in size and shape, she walks with a decided stoop, caused by the extremely high heels she has worn from babyhood.

"The Surprise," the "God of Fashion," the "Way to Beauty is Through the Purse," the "Esmerelda" and the "Land of Love" are dry goods stores kept by Frenchmen, and filled with the most expensive things ever exhibited to the public. While the "Red Sombrero" sells silk hats at three dollars to hundreds of dollars for sombreros covered with fifty pounds of silver and gold embroidery, the "Temptation," the "Reform," the "Flowers of April," the "Sun of May," the "Fifth of May," the "Christmas Night" and the "Dynamite" sell pulque at a laco a mug to the thirsty natives.

The names of the streets were such a source of unfailing interest to me that I cannot refrain from telling of some of the strangest and most peculiar ones. All the saints ever heard of or imagined are honored. The Mexicans do not say street after a name, in our fashion, but always say the street of – such as the Street of the Little Hand, of the Masons, of Montezuma, of the Magnolia Tree, of the Moon, of Grace, of Joy, of the Joint of God, of Jesus and Mother, of the Sad Indian, of Independence, of Providence, of Enjoyment, of the Hens, of the Steers, of the Slave, of Pain, of the Devil, of the Delicious, of the Dance, of the Green Cross, of the Crosses, of Cayote, of the Flowery Field, of the Cavalry, of the Chin, of the Heads, of a Good Sight, of a Good Death, of the Wood of the Most Holy Bench, of Christ's Mother's Prayer, of the Arts, of the Trees, of the Angles, Street of Mirth, Street of Bitterness, Street of the Love of God. Street of the Golden Eagle, of the Little Bird, of the Palm, of Progress, Street of Spring, Street of Papers, of the Lost Child, of Mosquitoes, of Paper Money, of Monstrosities, of Death, of the Wars, of Intense Misery, of the Mill, of the Barber Shop, of the Mice, of the Refuge, of the Clock, of the Kings, of the Rose, of the Queen, of the Seven Principals, of the Solitude of the Holy Cross, of the Soldiers, of the Hat, of the Vegetables, of Triumphs, of a Sot, of a Bull, of the Shutting up of Jesus, of the Shutting up of Money, of the Blind, of the Heart of Jesus, of the Body of Christ, Back of St. Andrews, Back of the Son of God, Back of St. John of God, Back of the Holy Ghost, Back of the Flowers, Back of the Flesh, Back of the Fruit; then there is the Bridge of the Little Cars, Bridge of the Raven, Bridge of the Holy Ghost, Bridge of Iron, Bridge of Firewood, Bridge of Mercy, Bridge of Jesus, and many others equally curious.

There are eleven streets named after Humboldt in the City of Mexico. Curious legends are attached to many of the streets, but many have been forgotten; the street which faces the National Palace, called Don Juan Manuel, is very interesting from its story, which, they say, is every word true. As we have no power with which to test its veracity it must pass without questioning. Here it is:

When the Spaniards first settled in Mexico there was one man named Don Juan Manuel, who, although blessed with a handsome wife, was always discontented and complaining because his family did not increase; this melancholy affected his digestive organs, until he became a victim of dyspepsia, which we all know leads to various whims and fancies. At any rate, he became possessed of the idea that his wife was unfaithful to his fitful and fretful devotion, and he sat up at night brooding over this, and writing down beautiful names he would hear and read of, that would be handy in case of any sudden and unexpected event whereby they could be utilized.

One night while thus occupied the devil appeared and told him to bring his nephew from Spain, and also to stand, wrapped in a long black cape, such as is yet worn by his countrymen, in front of his house at eleven o'clock that night (a very late hour for a Spaniard to be abroad in Mexico). The first man who passed would be the one who had stolen his wife's love, whispered the devil, and Don Juan Manuel must say to him: "My friend, what is the hour?" and, on the man's replying, continue: "You are a happy man; you know the hour of your death," then stab him to the heart. This done, he was to immediately feel relieved. His wife's love would return, and he would ever after be supremely happy.

The don, much elated at the promised downfall of an imaginary rival, and the ease it would bring to his worried mind, hastened to do the devil's bidding; the very next night, wrapped in his long cloak, he stood in the shadow of his house; just as the watchman's whistle, calling the hour of eleven, had ceased to sound way off in the distance, a man, as the devil predicted, came walking by. "My friend, what is the hour?" cried Don Juan Manuel. True to the historic courtesy of his birth, the stranger politely stopped and replied: "With your permission, eleven o'clock, Senor Don." "You are a happy man; you know the hour of your death," and the unsuspecting stranger fell, stabbed to the heart, while Don Manuel hastened into his casa.

But he found no relief. While he had no regret for the deed, his jealousy seemed to burn with increased fire: so the devil came again and told him he had killed the wrong man, but he must persevere – go out again, kill the man that he should see at that hour, and at last he would find the right one; the people began to talk about a man being found every morning dead at the same spot and in the same manner. But Don Juan was one of their highest by birth and rearing and was above suspicion. Their superstition made them attribute the deaths to an invisible power, and no investigation was made.

In the meantime, Don Juan's dearly beloved nephew had arrived from Spain, and was not only warmly welcomed by him, but by his wife, who hoped the nephew might be the means of helping to bridge the chasm, which for months had steadily been increasing between herself and her husband. Night came on, and the don went out to perform his deadly business. A man clad like himself came along, and Don Juan approached with, "My friend, what is the hour?" "Eleven o'clock. Adois," briefly answered the one addressed. "You are a happy man; you know the hour of your death," and the dark-clad stranger sank with a slight moan, while the don fled to his dreary chambers.

Morning dawned, and a dead man, as usual, was found. Don Manuel met them carrying the body into his casa, heard the screams of his wife, and saw the rigid face of his beloved nephew, dead, and by his hand! He rushed to his father confessor, whom he had not visited for so long, and begged absolution. "Thou must first repent," said the father. "Repent, repent!" cried the wretched man; "I am racked with misery. Grant me absolution." "Prove thy repentance first," answered the father; "go and stand beneath the scaffolding in front of the official building when the bell and watchman tolls the hour for midnight. Prove thy repentance by doing that thrice, then come to me."

After the first trial he returned to the father, begging that absolution be granted, for devils had wounded his flesh and tortured him as he had stood beneath the scaffolding. "No, twice more must thou stand there," was the unrelenting reply, and once again he went. Morning brought him more dead than alive to the good father's side. His face wore the hue of death, his form was trembling, his eyes were glassy and his words wild. "I cannot endure the third night. Angels and devils alike surround me. My victims ask me, with their cold hands about my throat and glassy eyes staring into mine, to name the hour I want to die. My flesh is bruised where they burn and prick me. My head is sore from the frequent pulling of my hair. Grant me absolution; they have showed me the bottomless pit of hell, and I cannot return!"

The good father prayed long and earnestly with him, that the Almighty power would deal leniently with his many crimes, but commanded the trembling wretch to spend the third and final night beneath the scaffolding. Dawn came, but it brought no hopeful man for the promised absolution. They found him hanging on the scaffolding dead. Some say the angels took him away because he had suffered sufficiently for his sins. Others say the devils hung him because he tried to escape the toil he had willingly accepted. But he was dead. His story was made known, and because of the strangeness of it, this street was named after him, and I never traversed it while in Mexico but that I felt sorrow for the poor insane wretch as he stood three nights beneath the scaffolding on Don Juan Manuel.

CHAPTER XXVIII.
A MEXICAN PARLOR.

MOST readers will probably be interested to know how custom rules that a parlor shall be furnished "in Spanish" as we quaintly say in Mexico. For the knowledge that all are of a different tongue makes a rather queer impression and it is quite common for foreigners to remark: "Oh, they can't hear, they are Spanish." We even get to think they cannot see and that people laugh and babies cry "in Spanish."

A parlor, or *sala*, is found in every private Mexican house, but until within the last two years there was not a hotel in the republic that had a parlor. Boarders entertained their friends in their bedrooms – and this is yet considered quite the proper thing to do. Some of the hotels now advertise as *Americanos* on the strength of having a little parlor. Calling or visiting is quite uncommon, as there is no society, and little sociability outside their home doors, yet occasionally relatives call on one another; still I have been with cousins who accidentally met at church, and though they were the best of friends, living within a dozen squares of each other, they had not exchanged visits for three years; this is quite common. I know two sisters living within four squares of each other who have not been in each other's house for a year. I hardly think the reason is a lack of sociability or hospitality, as, once within the massive walls of their *casa*, the Spanish courtesy is readily exhibited; they are your servants, and their house is yours for the time being, but the main causes are the gradual decrease of their once princely fortunes, and their laziness; the latter I regard, from close observation, as the chief fault.

Yet with all their retired habits they retain the "custom" of former generations as to how their parlor must be arranged and visits paid and received, as strictly as though they were in the midst of an ultra-society circle; their customs, I have been informed, are thoroughly Spanish and are the only ones practiced both in Spain and Cuba.

The *sala* is always on the second floor, as none but servants occupy the ground or first floor, and it is generally the only room in the house which boasts of a carpet. In several cases I have seen the floors made of polished wood and marble tiling; the walls are beautifully frescoed in colors, and the ceiling, which is always very high, has a magnificent painting in the center, the subject invariably of angels or a group of scantily-clad females. In each corner there are round, brass-edged openings of about ten inches in circumference, which serve as ventilators and very often a double purpose by letting scorpions in on unwilling victims.

The windows are but glass doors opening out upon little iron-railed balconies shaded by awnings. Each window-shade is transparent, and as the light shines through, it not only fills the room with some beautiful delicate tints, but discloses a lovely Southern scene. Cobweby curtains of creamy white hang from brass poles, suspended at least a foot and a half from the window, forming in themselves little nooks which would be idolized by romantically inclined "spoons" and "spooners" of the States.

The Mexicans are all good judges of paintings and many are talented artists; they do not harrow up one's sensibilities with dollar daubs of blue-trees, lavender-tinted skies and a mammoth animal with horns and tail, standing on a white streak in the foreground, which (the animal) placed cross-wise, could stand on all fours and never touch water. Nor does one's eyes have to long for the waters of Lethe because of tea prizes and Mikado ornaments. But a selection of good oil paintings and French-plate mirrors, all framed in brass, grace their rooms.

The piano is almost universal and occupies some nook by itself; the furniture for the *sala* is always cushioned and is composed mainly of easy chairs; the sofa – the seat of honor – is placed against the wall beneath some large painting or mirror and a large rug is laid in front. Starting from either end are the easy chairs which form an unbroken circle around the sofa, all thus being able to face it without turning their backs on any one. Directly at the back of the chairs, or facing the sofa, is a round table with a "crazy" patchwork cover – which craze has invaded even that country – or a knitted scarf. Then it is actually littered with ornaments of every description, leaving no empty space; as an Englishman rather tersely remarked to me, "They look like a counter in a crowded pawn shop."

All the chairs, and the sofa, have crocheted tidies on the backs, arms and seat, each separate, and enough to madden a Talmage convert. You may rise up slowly with an Andersonian grace and first one female politely begs permission to remove one of her tidies from your hat; then they will file into the next room, one by one, to see how *La Americanos'* sombrero becomes them, while another removes a white, delicately constructed thing from your "tournure" (what they dote on), which latter they have been dying to closely inspect, and to find how you manage to have it hang so prettily. And after you remove another tidy which has become fastened to your heel (although you can't imagine how), you detach yet another from the side trimmings of your dress. By that time, you are flustered, forget the Andersonian grace, and utter some emphatic words about tidies and tidy matters in general, and sit down with a real Castletonian kick.

The *sala* is not complete without at least two cabinets to hold the overflow of the center table. In all the odd corners are pedestals on which are statuettes in marble, bronze, or plaster-of-Paris, just as the owner's purse permits. Tropical plants in quaint jars of Indian design and construction and rustic stands are grouped about, and parrots, mocking-birds, and gayly-colored birds of high and low voices complete the attractions of the beautiful Mexican *sala*.

CHAPTER XXIX.
LOVE AND COURTSHIP IN MEXICO.

> "Why the world are all thinking about it,
> And as for myself I can swear,
> If I fancied that Heaven were without it,
> I'd scarce feel a wish to be there." *Moore.*

BENEATH the Mexican skies, where everybody treats life as if it were one long holiday, they love with a passion as fervent as their Southern sun, but – on one side at least – as brilliant and transient as a shooting star. Yet there is a fascination about it which makes the American love very insipid in comparison.

In childhood, boys and girls are never permitted to be together. There is no rather sweet remembrance of when we first began to love, or having to stand with our face in the corner for passing "love letters," or the fun of playing "Copenhagen" when we didn't run one bit hard. It is only of a dirty little schoolroom filled with dusky *ninos*, all of the same wearing apparel, who studied "out loud;" a fat little teacher who never wore tight dresses, and who only combed her hair "after the senoritas had gone home." A scolding French master and an equally bad music master completes the memories.

When Mexican damsels reach that "hood" which permits of long dresses and big bustles, they are in feverish expectation until, during a walk or drive, a flash from a pair of soft, black eyes tells its tale and a pair of starry ones sends back a swift reply, and with a tender sigh she realizes she has learned that which comes into the lives of them all. That night she peeps from behind her curtains and watches him promenade the opposite sidewalk back and forth, the gaslight throwing his shadow many feet in advance, which, she vows – next to him – is the most beautiful thing she ever gazed upon. She does not show herself the first time or does he expect it. Modesty or custom prevents. Just as he takes off his hat to breathe a farewell to her balcony, a white handkerchief flutters forth for an instant, he kisses his fingertips, the light goes out, and both retire, longing for *manana noche.*

Time goes on, and she gets bold enough to stand on the balcony, in full glare of the laughing moon, whilst he walks just beneath her. When it rains he will stand there until hat and clothing are ruined, to show his devotion. When she goes for a walk he is sure to follow slowly behind, and if chance offers he touches his hat slightly, and she with upraised hand deftly gives the pretty Mexican salutation. When the novelty wears off all this, she gets a pencil, paper, and cord, with which she transfers to him those sweet, soft little nothings which the love-stricken are so fond of, and the fair fisheress never draws in an empty line; hers are but the repetition of what almost any love-sick maiden would pen – badly written and mis-spelled, it is true; his is something of this style:

"BEAUTIFUL, ENTRANCING ANGEL, – Your loving slave has been made to feel the bliss of heaven by your gracious and pleasing condescension to notice his maddening devotion for you. I long to touch your exquisite hand that I may be made to realize my happiness is earthly. Life has lost all charms for me except beneath your fortunate balcony which has the honor of your presence. Only bless me with a smile and I am forever your most devoted, who lives only to promote your happiness.

"YOUR SERVANT WHO BENDS TO KISS YOUR HAND."

Every letter ends with this last, as we end ours "Respectfully." If they do not care to write it out fully they put only the initials for every word. If a girl is inclined to flirt she may have several "bears," but her fingers tell a different hour for each. If two should meet they inquire the other's mission, and their hot blood leads them into a duel – which, however, is less frequent of late years. No difference how much a girl may care for a duelist, she does not see him after he has fought for her.

Winter comes at last, and with it the annual receptions of the different clubs. A mutual understanding and many fond hearts beat in anticipation of the event. Once there they forget the eyes of their chaperons, and in their adorers' arms they dance the Spanish love-dance. It is really *the danza*. At all receptions it comes in every other dance and is played twice the length of any. It is the one moment of a Mexican's life, and I assure you they improve it. The danza is rather peculiar, and not at all pleasing to an *Americana*. It is nearly the waltz step reduced to a slow, graceful motion; the high heels and tight boots prevent any swift movement; the gentleman takes the lady in his arms and she does likewise with him – as nearly as possible – and in this way they dance about three minutes, then encircling, as two loving schoolgirls walk along, they advance, and, clasping hands with the nearest couple, the four dance together for a little while and then separate; this repeated by the hour constitutes the Spanish danza. Uninterrupted conversation is held continually as the girl's cheek rests against the gentleman's shoulder. Love is whispered, proposals are made, and arrangements for future actions perfected.

When parents notice a "bear," if they are favorably inclined, they invite him in, where he can see the object of his adoration hemmed in on either side by petticoats of forbidding aspect. When he once enters the house, it means that he has been accepted as the girl's husband, and there is no "backing out." The father sets a time for a private interview and when he calls they settle all business points: As to what the daughter receives at the father's death, when the marriage shall take place, where the bride is to live and how much the intended husband has to support her; the lawyer finishes all arrangements and escorts the engaged pair to a magistrate, where a civil marriage is performed – that their children may be legal heirs to their property. Even after this they are not permitted to be alone together; the intended bridegroom buys all the wedding outfit, for the bride is not allowed to take even a collar from what her father bought for her before.

The final ceremony is performed in a church by a *padre*, who sprinkles the young couple with holy water and hands an engagement ring to the groom, which he puts on the little finger of his bride, then the *padre* puts a marriage ring on both the bride and groom. After which, holding on to the priest's vestments, they proceed to the altar, where they kneel while he puts a lace scarf around their shoulders and a silver chain over their heads; symbolic that they are bound together irrevocably, as there is no such thing as divorce in Mexico. After mass is said the marriage festivities take place and last as long as the husband cares to pay for them, anywhere from three days to a month, and then, like the last scene on the stage, the curtain goes down, lights are put out, and you see no more of the actors who pleased your fancy for a short time.

The husband puts his wife in his home, which is henceforth the extent of her life. She is devoted, tender, and true, as she has been taught. She expects nothing except to see that the servants attend to the children and household matters – and she gets only what she expects. He finds divers amusements, for, according to the customs of his country, his "illusion" (what they call love) dies after a few days spent alone with his bride, and he only returns at stated intervals to fondle or whip his captive – just as fancy dictates. The men discuss at the club the fact that he has more loves than one, but they all have, and it excites no censure. But the world can never know what the bride thinks; private affairs are never made public. He can even kill her, as did their predecessor Cortes, and it will excite little or no comment. When matured years come on, she loses what good looks she had; three hundred pounds is nothing for weight, and on her lip grows a heavy, black mustache. She cares for nothing but sleeping, eating, drinking, and smoking the perpetual cigarette. And in this way, ends the fair Mexican's brief dream of the *grande passion.*

CHAPTER XXX.
SCENES WITHIN MEXICAN HOMES.

THE City of Mexico makes many bright promises for the future. As a winter resort, as a summer resort, a city for men to accumulate fortunes, a paradise for students, for artists; a rich field for the hunter of the curious, the beautiful and the rare, its bright future is not far distant. Already its wonders are related to the enterprising people of the States, who are making tours through the land that held cities even at the time of the discovery of America.

The Mexican Central road, although completed only five years ago, offers every, and even more, comforts than old established eastern roads. Many excursionists have had delightful visits here, and at present a number of Quakers have come to see for themselves what Mexico offers. One of the party was quizzing Mr. Theo. Gestefeld, editor of the *Two Republics*, on the advisability of opening a mission for the poor and degraded of Mexico. Mr. Gestefeld is a first-class newspaper man, formerly employed on the Chicago *Tribune*, and has a practical and common-sense way of viewing things. His reply should be studied by all coming to Mexico to stay. He said: "Their religion has been the people's faith always, even before Americans lived. They are fanatics, and trying to change or convert them is wasting time. Let their faith alone, and go out and buy a farm on the table lands and teach them how to farm and how to live. You will find them ready, willing, even anxious to learn. They will quickly imitate any way they know is better than theirs." The Quaker is still here, but, so far as known, has neither started a mission nor bought a farm.

Mexico is colder these last few days than the traditional oldest inhabitant ever remembered, but it is a pleasant change to the visitors who have left the snowbound country, even if a fire is an unheard-of thing.

People who read history form wrong ideas of how Mexican houses are built. They are square, plastered outside and decorated. Many are three and four stories in height. The windows, which are always curtained, are finished with iron balconies. Massive doors, on which are ponderous knockers of antique shape and size, keep from view the inhabitants of the Casa. A knock, and the doors swing open and a brown portero, dressed in the garb of his country, sombrero, serape and all, admits you to the lower court, where the stables are kept and the servants live. Beautiful flowers, rare orchids, and tall, waving palms are growing in rich profusion. Directly up through the center is a large, open square; a stairway, decorated in the highest style of art, leads to the different departments. Fine statuary, singing birds and fountains mingling with the flowers aid in making the scene superb.

Just the opposite of the States, the higher up a room is the better it is considered, and in hotels they charge accordingly, $1 first floor; $2 second; $3 third, and so on. A room is not healthy unless the sun shines into it; and they have no windows – just glass doors.

All the hotels in Mexico are run on the European plan. They have restaurants attached where the waiters, as long as they smile, cannot do too much for their customers. Mexico has several good hotels, of their kind, and most of them equal, if they are not superior, to the Iturbide – pronounced Eeturbeda – but Americans who run after royalty want to stop here so they can say they have stayed at the house which was the palace of the first emperor after Mexico was independent.

Mexico looks the same all over, every white street terminates at the foot of a snow-capped mountain, look which way you will; the streets are named very strangely, one straight street having half a dozen names. Each square has a different name, or designated as First San Francisco; the next block Second San Francisco. Policemen stand in the middle of the street all over the city, reminding one of so many posts. They wear white caps with numbers on, blue suits, nickel buttons. A mace now takes the place of the sword of former days. At night they don an overcoat and hood, which makes them look just like the pictures of veiled knights. Their red lanterns are left in the place they occupied during the daytime, while they retire to some doorway where, it is said, they sleep as soundly as their brethren in the States. At intervals they blow a whistle like those used by street car drivers, which are answered by those on the next posts; thus, they know all is well. In small towns they call out the time of night, ending up with tiempo sereno (all serene), from which the Mexican youth, with some mischievous Yankeeism, have nicknamed them Sereno.

It is very easy for those unaccompanied and not speaking Spanish to get around in Mexico. A baggage man meets the train out from the city, who not only attends to his regular duties, but gives any information regarding hotels that visitors may want. Numerous carriages of all kinds and descriptions, stand around the depot. Each one is decorated with a flag, by which the visitor may know the price without asking. White, red, and blue: fifty cents, seventy-five cents, and one dollar. The drivers often try to get the best of a tourist, especially if he speaks Spanish, and charge him one dollar for a seventy-five cent carriage. The Mexicans do not differ much from the Yankee hackman. If any, it is in favor of the Mexican. They do not cheat so much, because they are not sharp enough.

Pulque shops, where they deal out the national drink, are quite plenty. These are the only buildings in the city that are decorated. They are generally corner buildings, and the two sides have finely-painted pictures of ladies, ballet-girls, men on gayly-caparisoned horses, angels floating on clouds, etc. Numerous flags of black and red, or red and white, answer for a sign, but it is against the law to use the national flag. These saloons, or shops, as they are called, stand wide open, with no screens to hide the dirty bar and drinkers from the eyes of pedestrians. They are patronized by men, women, and children, and are kept open all the time.

"Sabe que es pulque –
Licor divino?
Lo beben los angeles
En vez de vino."

Know ye not pulque –
That liquor divine?

Angels in heaven
Prefer it to wine.

Pulque is the fermented juice of the agave, or so-called century plant, which matures in from five to fifteen years, instead of one hundred as generally believed. It grows wild here, but large plantations of it are cultivated. Just before the plant is ready to blossom the natives gather the big fat leaves together, around the bud, forming a sort of basin. The bud is then cut out and the juice from the stalk collects in the leaf-formed basin. One stalk will yield as high as two gallons a day for six months.

The pulque is collected in jars that the gatherers carry suspended from their shoulders. It is sucked out of the basin through a hollow bamboo or reed, and squirted from the mouth into the jar. A knowledge of this fact does not render the stuff any more palatable to foreigners. It is awfully nasty stuff, but they say that when you get acquainted with it you like it real well.

Mescal is a sort of brandy distilled from pulque, and will paralyze almost as promptly as a stroke of lightning. Metheglin – honey and water – is made from the honey ant; they are placed in a piece of bolting cloth and the honey squeezed out of them.

The street-car system here is quite unique. But first a few statistics may prove interesting; they run on ninety miles of rails, and carried last year nine million passengers; the company owns one thousand five hundred mules and horses, one hundred and thirty-nine first-class coaches, sixty-five second-class, forty-six platform or freight cars, and twenty-six funeral cars. They pay an annual dividend of six per cent, on a capital of $5,000,000. The Chairman of the Board of Directors, Senor Castillo, speaks Spanish and English; they are very particular about free passes, and so far, this year, have only issued six.

First-class cars are exactly like those in the States, and the second-class look just like the "Black Maria," except the wheels. Cars, just like open freight or truck cars on railroads, are used for hauling instead of wagons, and a dozen of these, loaded with merchandize, are drawn by one team. Movings and everything are hauled in this manner; the price charged is comparatively small. Cars do not run singly, but in groups of four and five. Even on the first-class cars men smoke as much as they wish, and if the women find it unbearable they go out and stand on the platform; there are two conductors on each car; one sells the tickets, the other collects them.

When the line was first opened an enterprising stockholder bought up all the hearses in the city and had funeral cars made. The coffin is laid on one draped car; white for young and black for old, and the mourners and friends follow in street cars hired for the purpose. A stylish funeral will have a dozen or more cars, the windows of which are hung with white crepe, and the doors with black; the drivers and conductors appear in black suits and high, silk hats; the horses are draped, and have black and white plumes on their heads. The cost of funerals ranges from $20 to $1500. A stylish one is a beautiful sight; the poor, by making application to the police, are given the funeral car and passage for two persons free; the low and poverty-stricken class also hire the coffins, and when they reach the cemetery the corpse is taken out, wrapped in a *serape* and consigned to a hired grave – that is, they buy the grave for five years, at the end of which time the bones are lifted and thrown in some corner, exposed to the gaze of the public, in order to make room for new-comers, and the tombstones – then useless – are laid in one heap by the gate. The people are no respecters of human bones; Americans always want to go back to the States to die.

Street car drivers, of which there are two on each car, are compelled by law to blow a horn at every crossing to warn pedestrians of their coming; the horns are similar, in tone and shape, to those used by fish peddlers in the States. Drivers of every kind of vehicles use the long lash whip of plaited leather exclusively, and they ply them quite vigorously on their animals; they also urge them to faster speed by a sound similar to that which the villain on the stage makes as he creeps upon intended victims when asleep, with his finger on his lips. It sounds like a whip lash cutting through the air. The carts in use here are of the most ancient shape and style; two large, wooden wheels support a big square box. One mule is hitched next to the wagon, and three abreast in front of that, and one still ahead; the harness baffles description. Drivers very seldom ride, but trot along beside their team with rope lines in their hands; they can trot at the speed of the mules with apparent comfort.

Mexico does not breakfast. When people go into the restaurants and order a breakfast the waiters look at them in wonder, and inform them in the most polite terms in the world that they have but coffee and dry bread for breakfast. It is asserted that to eat breakfast will cause a heaviness and dullness for the entire day, but whether this is true or otherwise, it cannot be stated, for since our arrival in Mexico we have been unable to find any other than as before mentioned – and black coffee at that. Every family takes their coffee in their bedrooms. It takes at least two hours to get through an ordinary dinner.

A description of dinner in a private family will, no doubt, prove interesting to most readers, especially if they understand the difficulty of obtaining admission into a family. A Mexican will be all politeness, will do anything for you, will place his house at your service, but he and his family will move out. He will do anything but admit you to the secrecy of his house. So, this experience is rare.

Dinner was announced and the gentlemen, in the most courteous manner, offered their arms, and we walked along the balcony to the dining-room. The lace-hung doors were swung open, and there before us was the table with plate, knife and fork, and a penny loaf of bread at each place. We sit down, take our napkins, and the waiters – always men – fill our glasses from the elegant water bottles that grace each end of the table. One dish, containing, perhaps, cold meat, salad, red pepper, radishes, and pickled beans, is served on plates, and the first ones taken away from us, although not used. After endeavoring to swallow some of this nauseating stuff, which the natives devour with relish, the servant removes the dish, our plates, knives and forks, and another equally strange and equally detestable dish is brought on. Thus, the feast continues, meanwhile breaking the penny loaf in bits and eating without a spread.

Butter, which commands $1 a pound, is never seen from one year's end to another, and jelly is an unheard-of dish. The last dish, and one that is never omitted from dinner or supper, is frijoles – pronounced free-holies – consists of beans, brown ones, with a sort of gravy over them. If a Bostonian were but to visit this country his intellectual stomach, or appetite, would be sated for once. Sliced orange, covered with sugar and cinnamon, is dessert, after which comes chocolate or coffee; the former superb, the latter miserable. With the coffee the ladies and gentlemen smoke their cigarettes.

Children are really good here, their reverence for their parents being something beautiful. When entering the dining room each one kisses its mother's hand, and when she asks them if they wish such and such to eat they reply: "With your permission." Although all are smokers they could not be persuaded to take a cigarette in their mother's presence. The pulque, which is also given around with the coffee, they refuse through respect to their mother; but they drink when she is not by, and of course she is aware of the fact, and has no desire to prohibit them from it. It is just their form of respect to refrain in her presence. A Mexican could not be compelled to eat of two different dishes from one plate. Even the smallest child is proof against persuasion on this point.

The frijoles, or beans, are served on a tortilla, a sort of corn-cake baked in the shape of a buckwheat cake. Another tortilla is folded together, and answers for a spoon. After finishing the beans, it is not considered proper or polite unless you eat your spoon and plate.

Every family has at least half a dozen servants. They are considered excellent when they receive five dollars a month, and board themselves. Sometimes they are paid three dollars a month, and allowed six cents a day to furnish what they want to eat. This sum is called the retainer. Women do the cooking, and the men wait on the tables, make the beds and nurse the babies. Contrary to the usual report, they are very, very cleanly. Every room in the house is swept daily; balconies and uncarpeted rooms scrubbed as often. Beds, which are in hospitals, have board or iron bottoms, and the hardest of hard pillows.

Brooms are an unseen article, notwithstanding the country furnishes the most beautiful broom corn in the world. It is bought in bunches and tied to a short stick, and used in that manner, forcing the sweeper to bend nearly double. Scrub brushes are but a bunch of coarse straw tied around the top with a string, but they make the floors perfectly white. There is a fortune here awaiting some lively fellow who will bring machinery and make brooms and brushes for the natives; the straw costs comparatively nothing, and is of the very best quality.

Lotteries swarm here, and are a curse to the poor. Men, women, and children sell the tickets along the streets, and the poor have such a mania for buying that they will pawn their clothing in order to obtain a ticket.

There are no newsboys in this country. Occasionally a boy is seen with a package of papers, but he does not call out like they do in the States. Women generally sell papers, which they fold and hold out toward passers-by, never saying a word.

The people appear just the opposite of lazy. They move along the streets with a trot, equal in speed to the burro; they never turn their heads to gaze at a stranger, but go along intent on their own affairs as if they realized the value of time and shortness of life.

Ladies in the States should import their servants from Mexico. Their hire is a very little sum: they furnish their own food; they are the most polite, most obedient people alive, and are faithful. Their only fault – and a very common one with servants – is that they are slow, but not extremely so. To children they are most devoted; as nurses they are unexcelled; their love for children amounts to a passion, a mania. As a common thing here, a girl of thirteen is not happy unless she has a baby; but with all that they are most generous with them. Much amusement was caused the other day by an American asking a pretty little black-eyed girl if the bouncing babe tied to her back was hers. "Si, senor, and yours, too," she replied, politely.

The men share the troubles of nursing with the women, and the babies, tied on their mother's or father's back, seem as content as if they were rocked in downy cradles. Babies, as soon as born, are clad in pantaloons and loose waist, irrespective of sex. There are no three-yard skirts on them. Boys retain this garb, but girls, when able to walk, are wrapped twice around the body with a straight cloth which serves for skirts.

If you ask a native in regard to the sex of a baby he will not say it is a boy or it is a girl, but "el hombre" (a man) or "la mujer" (the woman.) All efforts fail to make them say "hijo" (son) or "hija" (daughter).

CHAPTER XXXI.
THE ROMANCE OF THE MEXICAN PULQUE.

THE maguey plant is put to as many uses by the Mexicans as the cocoa palm is by the South Sea Islanders. All around Mexico, even on the barren plains where nothing else can exist, it grows in abundance. Its leaves are ten and more feet in length, a foot in breadth and about eight inches thick. Of course, there are smaller and larger growths, according to their age. After collecting strength for about seven years it sprouts from the center a giant flower stalk, twenty or thirty feet high, on which often cluster three thousand flowers of a greenish yellow color. These wonderful plants in bloom along the plains form one of the most magnificent sights in Mexico. At the very least, forty have been seen at one place, each vieing with the other to put forth the most beauty.

A prince named Papautzin, of the noble blood of the Toltec, discovered some fluid in a plant whose flowering spike had been accidentally broken off. After saving it for some time, he had the curiosity to taste it, and that taste was not only delicious to him, but was destined to moisten the throat and muddle the brain of the Mexicans for generations and generations, and to cause the curious and ever inquiring tourist to do like the whale did at the taste of Jonah. This noble prince was not like an Eastern Yankee; he did not keep his mouth shut until he obtained a patent. If he had, telephones and gas wells would be nowhere in comparison as a money-making scheme. He kindly sent some to his sovereign by his beautiful daughter, Xochitl, the flower of Tollan. The noble king drank and looked, looked and drank – the more he drank the more he liked the stuff; the more he looked the more he liked the girl. So, he kept her, a willing prisoner, and their son was placed upon the throne.

Generations after generations rolled by lovely Xochitl. The king, their son, and the illustrious discoverer had solved the wonderful problem. The maguey plant was cultivated by thousands, and oceans of its fluid had gone down the throats of the natives. This was the origin of the Mexican national drink, pulque. No estimate can be formed of the amount used, but it is enormous. It is simply water for the natives, and a pulque shop graces, almost invariably, every corner in the cities. As stated in a former chapter, these shops are the finest decorated places in Mexico. Superb paintings of all scenes grace the interior and exterior; flags float gracefully over the doors, and customers are always plenty. Men, women, and children can be seen constantly drinking from clay pitchers of a generous size, for the full of which they pay but two cents. No respectable Mexican would enter a pulque shop, but they all drink it at every meal.

The maguey is planted at the interval of three yards apart, and in such a manner that every way you look across an estate the plants run in a straight line; they thrive in almost any soil, and after planting need no more attention until the time of flowering, which is anywhere from six to ten years. The Indians know by infallible signs just when the flowering stem will appear, and at that time they cut out the whole heart, leaving only a thick outside, which forms a natural basin. Into this the sap continually oozes, and it is removed twice, sometimes thrice a day by a peon, who sucks it into his mouth and then ejects it into the jar he carries on his back. As soon as the plant exhausts all this sap, which was originally intended to give strength and life to the flowering stem, it dies, and is replaced by innumerable suckers from the old root. Great care must be exercised in cutting the plant – if the least too soon or too late, it is the death of it.

When first extracted the sap is extremely sweet, from which it derives its name, aguamiel (honey water). Some of this is fermented for fifteen and twenty-five days, when it is called madre pulque (the mother of pulque). This is distributed in very small quantities among different pigskins; then the fresh is poured on it, and in twenty-four hours it is ready for sale. Plants ready to cut are valued at about $5, but an established maguey ground will produce a revenue of $10,000 to $15,000 per annum. Pulque is brought to town in pig and goat skins. It has a peculiar sour-milkish taste, and smells exactly like hop yeast.

From the mild pulque is distilled a rum called mescal. It is of a lovely brown, golden color, and very pleasant to the taste. One can drink it all night, be as drunk as a lord, and have no big head in the morning. If it was once introduced into the States nothing else would be used, for no difference how much is drank, the head is as clear and bright as the teetotaler's in the morning. Nor is this the only use of the plant. Poor people roof their huts with the leaves, placing one on the other like shingles. The hollowed leaf serves as a trough for conducting the water. The sharp thorns are stripped off, leaving the fibers attached, and the natives use them as a needle, already threaded. Paper is made from the pulp of the leaves, and twine and thread from their fibers. The twine is woven into rugs, mats, sacks, ropes, harness, even to the bits, and dainty little purses, which tourists buy up like precious articles.

The wonderful productive powers of this plant do not end here. The expensive cochineal bug, used for coloring purposes and for paint, counts this maguey its foster-mother. On its wide leaves does it live externally and internally until the gatherer comes and plucks it off, probably to color some dainty maid's gown in the far distant land or tint some sky of an artist's dream.

Yet maguey thinks it has not done enough for mortals, and it accomplishes one more thing for which the Mexicans would treasure its memory but Americans would gladly excuse it. Clinging to the shadiest side, in a childlike confidence, is a long green worm, similar to the unkillable cabbage worm of the States. Peons in a gentle manner, so as not to crush or hurt, pluck these tender young things, and, putting them in a vessel, bring the fruits of their work to town. Nothing can be compared to the way and haste in which people buy them. Fried in butter, a little brown milk gravy around, and they are set on the table as the greatest delicacy of all Mexican dishes. It is needless to add that the natives eat them with wonderful relish, and are quick to say, "We know what these dainty things are, but you folks eat oysters!"

CHAPTER XXXII.
MEXICAN MANNERS

AMONG the most interesting things in Mexico are the customs followed by the people, which are quaint, and, in many cases, pretty and pleasing. Mexican politeness, while not always sincere, is vastly more agreeable than the courtesy current among Americans. Their pleasing manners seem to be inborn, yet the Mexican of Spanish descent cannot excel the Indian in courtesy, who, though ignorant, unable to read or write, could teach politeness to a Chesterfield. The moment they are addressed their hat is in hand. If they wish to pass they first beg your permission, even a child, when learning to talk is the perfection of courtesy. If you ask one its name it will tell you, and immediately add, "I am your servant" or "Your servant to command." This grows with them, and when past childhood they are as near perfection in this line as it is possible to be.

When woman meets woman then doesn't come "the tug of war," but instead the "hug and kissing;" the kissing is never on the lips, but while one kisses a friend on the right cheek, she is being kissed on the left, and then they change off and kiss the other side. Both sides must be kissed; this is repeated according to the familiarity existing between them, but never on the lips, although with an introduction the lips are touched. The hug – well, it is given in the same place as it is in other countries, and in a right tight and wholly earnest manner. From the first moment they are expected to address each other only by their Christian names, the family name never being used.

The parlor furniture is arranged the same all over Mexico; the sofa is placed against the wall and the chairs form a circle around it; the visitor is given the sofa, which is the "seat of honor," and the family sit in the circle, the eldest nearest the sofa; the visitor expects to be asked to play the piano, which she does in fine style, and then the hostess must play after her or commit a breach of courtesy, which social crime she also commits if she neglects to ask the guest to play; visitors always stay half a day, and before leaving she is treated to a dish of fine dulce, a sweet dessert, cigarettes and wine; then mantillas are put on, blessings, good wishes, kisses and embraces are exchanged, each says "My house is yours; I am your servant," and depart. All the rules of decorum have been obeyed.

When men are introduced they clasp hands, not the way Americans do, but with thumbs interlocked, and embrace with the left arm; then the left hands are clasped and they embrace with the right arm, patting the back in a hearty manner; the more intimate they become the closer the embrace, and it is not unusual to see men kiss; these embraces are not saved for private or home use, but are as frequent on the streets as hat tipping is here; the hand clasping is both agreeable and hearty. They clasp hands every time they part, if it be only for an hour's duration, and again when they meet, and when careless Americans forget the rule they vote them very rude and ill-bred. Undoubtedly, as a nation, we are.

On the street a woman is not permitted to recognize a man first. She must wait until he lifts his shining silk hat; then she raises her hand until on a level with her face, turns the palm inward, with the fingers pointing toward the face, then holds the first and fourth fingers still, and moves the two center ones in a quick motion; the action is very pretty, and the picture of grace when done by a Mexican senora, but is inclined to deceive the green American, and lead him to believe it is a gesture calling him to her side. When two women walk along together the youngest is always given the inside of the pavement, or if the younger happens to be married, she gets the outside – they are quite strict about this; also, if a gentleman is with a mother and daughters, he must walk with the mother and the girls must walk before them. A woman who professes Christianity will not wear a hat or bonnet to church, but gracefully covers her head with a lace mantilla. No difference how nicely she is clad, she is not considered dressed in good taste unless powdered and painted, to the height reached only by chorus girls. Four years ago, the Americans tell me, the Mexican women promenaded the streets and parks and took drives in ball-dresses, low neck, sleeveless, and with enormous trains; this has almost been stopped, although the finest of dresses, vivid in color, and only suitable for house or reception wear, are yet worn on Sundays.

Everybody wears jewelry, not with good taste, but piled on recklessly. I have seen men with rings on every finger, always excepting the thumb; and the cologne used is something wonderful. You can smell it while they are a square off, and it is discernible when they are out of sight. A man is not considered fashionable unless he parts his hair in the middle, from his forehead to the nape of his neck, and dress it *a la* pompadour. The handkerchief is always carried folded in a square, and is used alternately to wipe his dainty little low-cut boots and the face. Afterward it is refolded and replaced in the pocket.

Visitors are always expected to call first to see their friends when in town, as it would be a great breach of decorum for a family to call on a visitor before he or she came to their house. If two or more people meet in a room and are not acquainted they must speak, but not shake hands; they can converse until someone comes, when they will accept an introduction and embrace, as if they had just that instant met. When one occupies a bench in the park with a stranger neither must depart without bidding the other farewell, and very often while murmuring adieus they clasp hands and lift hats.

Mexicans in talking employ a number of signs, which mean as much to them and are as plainly understood as English words would be to us. They speak their sign-language gracefully; indeed, they are a very graceful people, and yet they never study it or give it a thought. When they want a waiter in a restaurant, or a man on the streets, they never call or whistle, as we would do, but simply clap the hands several times and the wanted party comes. The system is very convenient, and far more pleasing than the American plan. When wishing to beckon any one, they throw the hand from them in the same manner as Americans do if they want anyone to move on. To go away, they hold the fingers together and move them toward the body.

They never say that a man is drunk; it sounds vulgar, and, as they will "get that way," they merely place the index finger on the temple and incline the head slightly toward the person meant. They could never be abrupt enough to say any one was crazy or had no brains, so they touch the forehead, between the eyebrows, with the first finger. To speak of money, they form a circle of the thumb and forefinger; to ask you to take a drink or tell the servants to bring one, the thumb is turned toward the mouth; to ask you to wait a little while, the first finger is held within a quarter of an inch of the thumb. To hold the palm upward, and slowly move the hand backward and forward, says as plain as English "I am going to whip my wife," or "I whip my wife." If they want you to play a game at cards, they close both fists and hold them tightly together. Touching the thumb rapidly with the four fingers closed means you have much or many of anything, like many friends. Making a scissors sign with the fist and second finger means you are cutting someone in the back. Whittling one forefinger with the other means "you got left." When courting on the balcony and the girl smooths her lip and chin, you are warned to get out; "the old man is coming." In company, when one is so unfortunate as to sneeze, they are greatly insulted, and the company is badly wanting in good manners unless, just as the sneeze is finished, everyone ejaculates "Jesu," "Jesucristo."

CHAPTER XXXIII.
NOCHE TRISTE TREE.

I PRESUME everybody who knows anything about history remembers reading how Cortez, when he thought he was going to lose the fight for Mexico, on July 10th, 1520, retired under a tree and wept.

Since that time the tree has been known to the inhabitants as the Noche Triste (the sad night). It stands before an ancient chapel, in a public square of the little village of Popotla. I don't know why, for I could never think of Cortez except as a thieving murderer, but the Noche Triste receives a great deal of attention from the natives and all the tourists. On the second of May 1872, the tree was found to be on fire. A citizen of Popotla, Senor Jose Maria Enriquez, who venerated the old relic, followed by hundreds of people, rushed to its rescue.

They did what they could with buckets, and at last two hand pumps were brought from an adjoining college. It is said that fully five thousand people visited the burning tree that day. After burning for twenty-four hours the flames were conquered. Since then Noche Triste has been inclosed by a high, iron fence; despite the fire it is yet a grand old tree.

Everybody visits the cathedral of Mexico. It is a grand old building, of enormous size, and covered with carved figures, facing the zocalo. It is surrounded by well-kept gardens, in which are many beautiful statues and ancient Aztec figures. In the cathedral is the tomb of the Emperor Yturbide, and superb paintings, some by Murillo. The history of the cathedral is interesting; it was the church of Santa Maria de la Asuncion until January 31, 1545, when it was declared the metropolitan cathedral of Mexico.

Philip II. issued a royal decree that the cathedral should correspond to the magnificence of the city, and in 1573 the work was begun. It occupies the very ground on which stood the principal temple of the Aztecs; the site was bought from the Franciscan monks for forty dollars.

A period of forty-two years was consumed in laying the foundations, raising the exterior walls, building the transverse walls of the chapels, working the columns to the height of the capitals, and making some progress upon the domes.

The architecture of this temple pertains to the Doric order. The structure is one hundred and thirty-three Spanish yards in length, and seventy-four in width. In has one hundred and seventy-four windows, and is divided into five naves, the principal one of which measures fifty-three feet in width from column to column. The aisles correspond in number to the thirty-three chapels, formed by twenty pillars, ten on each side; from base to capital the pillars measure fifty-four feet in height, and fourteen in circumference. The roof is composed of fifty-one domes or vaults, resting upon seventy-four arches. The church is pyramidal in form, its height diminishing in regular proportion from the main nave to the chapels. There are three entrance-doors on the southern front, two on the northern, and two on each of the sides.

After ninety-five years of continual work, the final solemn dedication was celebrated December 22, 1677.

The cost of the cathedral, exclusive of the external decoration, at least of the Sagrario, amounted to $1,752,000, so that it may well be said that two and a half million dollars were invested in the two churches, whose erection extended over more than a century.

During my six months in Mexico I received hundreds of letters from men asking my advice about their coming to Mexico for business purposes. I never give advice, but if I were a man and had a certain amount of patience I should go to Mexico. If one can get used to the people and their *manana* movements, the place is perfect, the land, in most localities, is the easiest imaginable to cultivate. A farmer can have as many harvests a year as he has space. He can sow in one place and harvest in another, so perfect is the climate. The only complaint is of the lack of water, but as it is always to be found six feet under the surface of the earth one can have it. Anything will grow if put in the ground. I visited one place that had been barren three years previous, and it was the most beautiful garden spot in Mexico. The trees were equal to any nine-year-old trees in the States. There is no weather to interfere with their growth.

A great number of Englishmen, Germans and French have settled in Mexico, and by their thrift are accumulating fortunes rapidly. Barring a little dislike, the Americans have the same chances.

Mexico produces better broom-corn than the United States, and for the smallest possible cost and trouble. Very few farmers interest themselves in broom-corn, so there is a place for Americans to step in and make money.

Silk culture could also be made one of Mexico's principal industries. It can be carried on with little or no capital. Anyone who possesses a few mulberry trees can, without abandoning his regular work, care for silkworms. An ounce of silk-worm eggs costs five dollars, and it will produce not less than fifty kilogrammes of cocoons that are worth one dollar per kilogramme.

It is only necessary to buy eggs the first time, for the worms keep producing them. The mulberry tree thrives in all parts of Mexico and the silk-worm needs no protection of any kind from the climate, nor are they subjected to diseases here which elsewhere cause great loss. It costs less to raise silk-worms in Mexico than in Europe, and a far better quality are produced. Mulberry shoots will produce sufficient foliage to maintain silkworms within three years after planting.

The eggs, while containing the embryo silk-worm, have a dull lavender color, but after discharging the worm they resembled little sugar pills. The worms were about one-sixteenth of an inch long, but the first week of moulting shows them to be half an inch long and the second week one inch. For the third moulting they are placed on perforated paper, through the holes of which the worms crawl. This relieves the attendant of considerable labor in transferring them. The fourth week the head is white, and the worm has attained its normal growth. There is nothing now for the worm to do but loaf around and lunch on mulberry leaves until the eighth or ninth day after shedding its skin the fourth time, when he or she, as the case may be, proceeds to form its cocoon. It is then of a golden transparent color. It takes about five days for the industrious worm to finish its cocoon. Then, to destroy the moth inside, it is subjected to heat, and the cocoon is then ready for spinning.

When ready for use the cocoons are soaked in a tub of water until all the glutinous substance is removed. With a small whisk-broom the cocoon is brushed until ends, which are as fine as a cobweb, come loose. They will then reel off without breaking. One cocoon will give four hundred yards of raw silk.

Indian rubber trees are also easily cultivated in Mexico, and the demand for them is large. It's easy to make a comfortable income in Mexico, if one goes about it rightly.

CHAPTER XXXIV.
LITTLE NOTES OF INTEREST.

SUPERSTITION is the ruin of Mexico. While we were there some children found a shell containing an image of the Virgin. The matter was deemed miraculous, and they directly decided to build a chapel on the spot where the shell was found.

In the State of Morelos exists a stone that they say was used before the conquest to call the people to labor or to war. The stone appears to be hewn. In the center of the upper part is a hole which runs into the heart of the stone, forming a spiral. On fitting to this a mouthpiece and blowing, the sound of a horn is produced, somewhat melancholy in tone, but so loud that it can be heard a great distance; the ranchmen of that locality employ it as a means of calling their flocks and the animals quickly obey the summons. It is known as the "Calling Stone."

There is a tradition about this stone; they say that no difference where it is taken, that by some invisible means it always goes back to the spot it has occupied for the past century. They say that once it was even chained in a cellar, but in the morning, it was missing, and when they searched for it, it was found in its old position.

Mexico abounds with the most beautiful and wonderful flowers. Many are unknown even to horticulturists. One of the novel flowers I heard of was one which grew on the San Jose hacienda, some twenty-two leagues from the City of Tehuantepec. In the morning it is white, at noon it is red, and at night it is blue. At noon it has a beautiful perfume, but at no other time. It grows on a tree.

There are very few fires in Mexico, and it is a blessing to the citizens; they have one fire company, but no alarms. When there is a fire the policemen nearest give the customary alarm, three shots in the air from his revolver; the next policeman does the same, and on up until they come to the policeman near the firemen's office. The fires are always out or the place reduced to ashes before these noble ladies put in an appearance.

On every corner is hung a sign, giving a list of all the business places on that block.

The turkeys in Mexico are the most obliging things I ever saw; they are brought into town in droves and they never scatter, but walk quietly along, obeying the voice of their driver. If he wants a drink he makes them lie down and they stay until he returns.

Mail is delivered every day in the week, Sunday not excepted. Every letter-box contains a slip which the carrier fixes, which tells when the next collection will be made. Printed slips are published daily, and hung in the corridors of the post-office, of unclaimed letters and papers, and of those that have not gone out for lack of postage.

Houses are never labeled "To Let" when they are empty; a piece of white paper is tied to the iron balcony and everybody knows what it means. No taxes are paid on empty houses or uncultivated land. People never rent houses by the year, but by the day or week; they can move at any time they wish; this makes landlords civil.

Grass is cut in the park with a small piece of zinc, which is sharpened on a stone, and it is raked with a twig broom.

No houses have bathrooms, but the city is well supplied with public swimming baths. One can have a room and private bath for twenty-five cents. Everybody of any note takes a bath every morning. It is quite a pretty and yet strange sight to see the beautiful young girls coming leisurely up the prominent thoroughfares early in the morning, with their exquisite hair hanging in tangled masses, often to their feet. They are always attended by a maid.

Mexican ladies have a contempt for people who do not have servants. They never carry anything on the streets; but always have a mozo, even to carry an umbrella.

Because Vera Cruz has such a large death rate from yellow fever the Mexicans have named it *La Cindad de los Muertos* (the city of the dead.)

In Yucatan the Maya language is still used. It is very musical and is written all in capitals.

It is considered polite and quite a compliment for a man to stare at a lady on the streets. I might add that the men, by this rule, are remarkably polite.

Families employ street musicians by the month, to visit them for a certain time daily. The hand-organs there are most musical instruments.

Shoes are never marked with a number, but are fitted until they please the buyer. The shoes worn on the street are what would be the pride of an actress. They are very cheap.

The easiest English word for the Mexican to learn is "all right." Even the Indians catch it quickly. They all like to speak English.

Butter is seldom seen in Mexico. The only way they have of getting it is by its forming from the rocking on the burro's back while being brought to town. It is skimmed off the milk by the hand and is sold at a big price. It is never salted. The butter is always wrapped in corn husks, looking exactly like an ear of corn until it is opened. They also make cottage-cheese, and tying it up in green reeds sell it. Salt is very expensive.

It costs a single man about one hundred and fifty dollars a month for his room rent and board. He must also retain the chamber-maid and the *patero* (door-keeper,) with certain amounts. Young men never carry night keys in Mexico, because they weigh about a pound.

According to law every door must be locked at ten o'clock, and all those entering afterward must pay the *patero* for unlocking and unbarring the heavy portals.

The poor, when dead, are carried to the graveyard on the heads of *cargadores*. If the coffin is only tied shut with a rope, it is borrowed for the occasion. The body is taken out at the cemetery and consigned, coffinless, to mother earth.

The Mexicans began to call the Americans *gringos* during the war. They say the way the title originated was this: at that time an old ballad, "Green grows the Rushes, O!" was very popular, and all the American soldiers were singing it. The Mexicans could only catch "green grows" and so they have ever since called the Americans "gringos."

Newspapers are published every day in the week except Monday. Sunday is always a feast day, and as no one will work then, the paper cannot be gotten out for Monday.

Mexicans never suffer from catarrh; they say it is because they will not wash the face while suffering from a cold. They say a green leaf pasted on the temple cures headaches.

The women in Mexico are gaining more freedom gradually; they have them now as telegraph and telephone operators. Some Mexican bachelors use the telephone for an alarm clock, that is, they have the girls wake them by means of the telephone placed in their room.

No bills are legal unless they are stamped. Every man has a peculiar mark which he scratches beneath his name. It is a sort of a trade mark, and makes his name legal.

The Indian women have some means of coloring cotton so that it will never fade.

There are public letter writers on the plazas, where one can have the correspondence attended to for a small sum.

Letter-writing is an expensive thing in Mexico; to all points not exceeding sixteen leagues, they pay ten cents for a quarter of an ounce, or fifty cents an ounce. Postal cards are two cents; to send a letter to the United States only costs five cents. Every state in Mexico has its own stamps.

Some haciendas are enormously large in Mexico. One man owns a farm through which the railroad runs for thirty miles. It is said to comprise ten thousand square miles.

The public schools in Mexico are similar to those in the States fifty years ago; the schools are never mixed; the boys attend one place and the girls another; the advanced teachers are elected, and are given a house to hold the school in, and one hundred dollars a month for conducting it. For the others they get a house somewhere, and from thirty to sixty dollars; ten years ago, girls were not taught spelling or writing in public schools; they are now taught all the common branches and English, which has replaced French; sketching, music, fancy-work, and plain sowing; the hours are from 8 to 12.30, and from 2 to 6; they are thoroughly taught the geography of their own country, but they absolutely learn nothing of other lands.

CHAPTER XXXV.
A FEW RECIPES FOR MEXICAN DISHES.

PROBABLY someone would like to make a few of the dishes most common to the Mexican table. Of course, you will think them horrible at first, but once you acquire the taste, American food is insipid in comparison.

Recipe for tortillas: – Soften corn in alkaline water, then grind it fine, pat into round cakes, and bake on a thin, iron pan. Eat while hot. They are made very good by wrapping them around meat, or a seasoned pepper.

Albondigas (meat balls): – Take equal parts of fresh pork and beef, say one pound, cut, as for sausage, put in salt, pepper, a small piece of soaked bread, and one egg, well beaten; make into small balls, putting in each a piece of hard-boiled egg, an almond and a raisin. In a dish of hot lard put five or six crushed tomatoes, a little chopped onion, salt, pepper, and broth. Let boil a few moments, and then put in the balls. When the meat is cooked it is ready for the table.

Rice with chicken or fresh pork: – Wash and dry the rice; have a dish of hot lard, put in the rice, fry a few moments, then add chopped tomatoes, onions, salt, pepper, two or three thinly sliced potatoes, and a few peas; cook a few moments, then pour into it the chicken or pork and some of the broth in which they have been boiled.

Stuffed red peppers: – Open the pepper, take out the seeds and wash and dry carefully. Boil and then chop fine as much fresh pork as you will need to stuff your peppers. In a dish of hot lard put the meat with plenty of fine-cut tomatoes and onions, salt and pepper. Boil a few moments, then add a little sugar, cloves, cinnamon, almonds, and raisins cut in half, cook a little, then fill the peppers. If you have eight peppers beat three eggs, whites and yolks separately; when well beaten put together, and in this roll the peppers, having first sprinkled over them a little flour. Have a dish of hot lard, to which has been added a little ground tomato, cinnamon, salt, pepper, and a little water. Boil a few moments, then put in the peppers, having first fried them in hot lard. Boil a few moments, and they are ready for use. The peppers can be filled with cheese if preferred, instead of meat.

Green peppers with eggs and cheese: – Roast the peppers over the coals, take off the thin skin, take out the seeds, wash and cut into thin strips. In a dish of hot lard put some tomatoes and onions, cut fine, and about two cups of water. When boiling, break in as many eggs as desired. When cooked, put in the peppers and slices of cheese. Rightly prepared, it is delicious.

Cocoanut dulce: – Grate fine two cocoanuts. Put in a dish three pounds of sugar, let boil, take off the scum, then add the cocoanut, stirring all the time. After a little a bowl of cream, then later eighteen eggs, well beaten. Let cook, stirring constantly, until, when you pass the spoon through the middle of the mixture, you can see the bottom of the dish; then take off. Put in platters. Peel and cut almonds in half; put them in as thickly as you please. Pass over it a hot iron until nicely smoothed.

Pineapple and sweet-potato dulce: – Grate pineapple, and boil sweet potatoes, half and half. For one pineapple two pounds of sugar; let boil and skim. Put in and boil, stirring all the time, until you can see the bottom of the pan as the spoon passes through the center.

Rice and almonds: – One ounce of grated almonds, one ounce of rice washed and ground; put in enough milk so it will pass through a cloth; put tins in a quart of milk, with three yokes of eggs and sugar to taste; boil until well done; flavor to taste.

CHAPTER XXVI
SOME MEXICAN LEGENDS.

THERE is hardly a spot in Mexico that has not some romantic history connected with it; and the tales are always so beautiful and full of thrilling romance. I would like to live in Mexico some time, and devote all my attention to gathering these interesting stories. I have given samples of them in the history of Don Juan Manuel.

The Street of the Jewel is also connected with a story full of love and its companion, despair. Here dwelled Gasper Villareal and his wife, Violante Armejo. Gasper was a man of moderate means, but he had enough to preserve his wife from labor. She was of wondrous beauty but quite strange, she only cared to hide herself in her convent-like home. She loved her husband, and he was as jealous as a Mexican can be.

One day a young noble. Diego de Fajardo, rode by the door, and, being thirsty, he asked the mozo for a drink. Violante sat in the corridor, looking upon the garden, and dreaming, doubtless, of her absent lord. True to the instincts of her race, she ordered the mozo to take the stranger a glass of wine. The servant did her bidding, explaining to the young cavalier the reason of the change in his refreshments. Diego de Fajardo felt that it would be churlish to ride away without acknowledging the gracious hospitality. He tossed his bridle to the man and passed into the garden.

Violante still sat in her hammock, garbed in spotless white, the perfection of beauty, grace and innocence. The young *caballero* had not uttered his thanks until he had vowed to win Gasper Villareal's lovely wife.

Day after day he watched the casa, waiting for an opportunity to find the wife alone. At last fate favored him. It was near nightfall when he saw the husband come forth, and, taking saddle, ride toward the city. In a moment, eager and confident, he fell on his knees before Violante and confessed his love.

She did not full into his arms, but she spurned him and with such anger that he saw his conduct in its true light, and, repentant he arose from his knees and left her. Violante started to her chamber to seek her rosary and to cool her throbbing brow with the touch of holy water, when her foot struck a sparkling object; it was a bracelet, With her name, "Violante," in diamonds, close beside the coronet and arms of De Fajardo.

As she stood her husband entered. Having to return for something, he had been struck with horror to see a man rush from his gateway. There stood his wife with the jewel in her hand, the evidence of her guilt. Without a word he sunk his dagger in her breast. As she sank lifeless to the floor, he snatched the gleaming bracelet from her stiffening fingers and left the house.

Diego de Fajardo was wakened in the morning by his mozo. Something had happened and he was wanted to go out in the street to see if he could understand it. Tremblingly he obeyed. On the pavement, Gasper Villareal lay rigid, his garments soaked with his life's blood. Near the bronze knocker of the massive door was a splendid diamond bracelet, suspended on a blood-stained dagger.

In 1550 the lake of Texcoco overflowed, and almost submerged the City of Mexico. Among the objects found drifting upon the water was a large canvas, on which appeared a beautiful representation of the Virgin. None could determine where it came from, so a chapel was built for it. It is called "Our Lady of the Angels." For centuries it has received the veneration of man.

Another inundation occurred in 1607, and all the chapel, except the side holding the Virgin's picture, was washed away. Despite all the storms the picture was said to be as bright as if just from the painter's brush. A new chapel was built around this marvelous painting, which stood until 1627, when another flood took it all away excepting the one wall holding the Virgin's likeness. There, neglected and unprotected, it stood as the storms had left it until 1745, when a succession of public calamities drove the people to implore the succor of the Virgin. A building was again erected around the uninjured painting. Thus, until the present day, the people in need seek the painting to pour forth their prayers at its feet.

El Desierto and its old Carmelite convent occupy the most charming spot in Mexico. It is only fifteen miles from the capital, and the way is along the most romantic and picturesque road a Southern clime can produce. The forest that surrounds El Desierto is composed of the largest trees in the valley, hardly excepting those of Chapultepec. The convent was a group of massive buildings, domes and turrets, now crumbling into decay. In 1625 the monks retreated to this wilderness to mortify the flesh, and strange stories of their serio-jovial life, their sparkling wines and romance of their hermit-like existence come creeping down through centuries; the jolly monks are no more, and the winds sigh through the mighty forest that has ridden romance, love and tragedy from the world.

The conqueror, Cortez, not satisfied with robbing the grand old Aztec king, Montezuma, of his land and life, also robbed him of his daughter. The poor woman, after he deserted her, died in a convent, leaving a daughter, the child of Cortez. This daughter of Cortez, and granddaughter of Montezuma, was married very young to a Spanish captain, Quinteros. There are now in Puebla descendants of that illegal love.

CHAPTER XXXVII.
PRINCESS JOSEFA DE YTURBIDE.

I CANNOT close this little book without speaking of one of the most remarkable and brilliant women in Mexico, the only daughter of the emperor. After the execution of the emperor the family came to the States, and settled in Philadelphia. Josefa was sent to Georgetown to receive an English education, and she yet retains a love for America and its people. When Maximilian entered Mexico, he restored the titles to the Yturbide family, and invited the cultured princess to become a member of his imperial household. Subsequently Emperor Maximilian adopted Augustin Yturbide, grandson of the late emperor, and appointed the Princess Josefa guardian of the "prince imperial." Maximilian soon recognized the wonderful executive abilities of the princess, and he consulted her on momentous occasions. Had he taken her advice, I doubt not but that Mexico would have had an empire to-day.

After the fall of Maximilian, Mrs. Yturbide (formerly Alice Green, of Washington, D. C.) claimed and recovered her son, who had been temporarily "heir presumptive" to the throne of Mexico. The Princess Josefa went to the court of Austria. Nine years ago, she returned to Mexico, where she lives in seclusion.

She is one of the loveliest women, in every respect, I ever met. Her rooms at the Hotel Humboldt are plain, but contain many little mementos of former glory. The pictures and busts of the unfortunate emperor and empress occupy prominent positions.

"Carlotta was only twenty-three years old when she came to Mexico," said the princess. "She was a beautiful girl, with a creamy complexion, dark eyes and hair. She worshiped her young husband, as he did her, and she was ambitious for his sake. What a sad fate was theirs!"

The princess then showed me five letters she had received from Carlotta, written in English, after the emperor's death; they gave no evidence of her insanity. The princess has never received any recompense for the land which the government took from her father, and even a pension due her, which now amounts to some hundred-thousands, has never been paid. She receives many promises from Diaz but never the money.

The worst things the Mexicans ever did for themselves was to shoot Maximilian. They have never had one quarter so good government since. They had sworn good faith to the emperor and said if he sent part of the French army back they would support him. He believed them, and when he found that they were dishonest he applied to Napoleon for aid. When he received no answer, the empress, eager to save her noble husband, started to beg Napoleon personally for help, much against the wish of Maximilian.

The republican powers getting too strong for the emperor, some advised him to seek refuge until things grew calmer. The refuge he sought was the prison they had prepared for him. He walked into it, and he never came forth until the day he was shot. His bosom friend, Lopez, whom the emperor had enriched, had made a general, and entrusted him with all his secrets, betrayed him to his enemies. On June 19, 1867, Maximilian and his brave comrades, Miramon and Mejia, were led forth to a little hilt near Queretaro and shot. Maximilian's last words were: "Poor Carlotta.'" Three little black crosses now mark the spot where those noble men died.

[THE END.]

Nellie Bly, 1888

Printed in Poland
by Amazon Fulfillment
Poland Sp. z o.o., Wrocław